Mastering RabbitMQ

Master the art of developing message-based applications with RabbitMQ

Emrah Ayanoglu

Yusuf Aytaş

Dotan Nahum

[PACKT] PUBLISHING open source community experience distilled

BIRMINGHAM - MUMBAI

Mastering RabbitMQ

First published: December 2015

Production reference: 1211215

Published by Packt Publishing Ltd.
Livery Place
35 Livery Street
Birmingham B3 2PB, UK.

ISBN 978-1-78398-152-6

www.packtpub.com

Credits

Authors
Emrah Ayanoglu
Yusuf Aytaş
Dotan Nahum

Reviewers
Steve Fosdal
Van Thoai Nguyen
Jorge Puente-Sarrín
Ken Taylor
Héctor Veiga

Commissioning Editor
Ashwin Nair

Acquisition Editor
Reshma Raman

Content Development Editor
Anish Sukumaran

Technical Editor
Abhishek R. Kotian

Copy Editor
Pranjali Chury

Project Coordinator
Mary Alex

Proofreader
Safis Editing

Indexer
Monica Ajmera Mehta

Graphics
Disha Haria

Production Coordinator
Conidon Miranda

Cover Work
Conidon Miranda

About the Authors

Emrah Ayanoglu has been into technology since a young age, when he was programming with his Tandy 1000 using Basic language. His deep interest and passion for programming lead to him pursue computer engineering at Bilkent University, Ankara. He now works as a software engineer and heavily works on integrating software systems using RabbitMQ.

He frequently speaks at different tech conferences about scalability and real-time web applications where RabbitMQ has a major role. Additionally, he participates in different open source projects.

For the future, he hopes to participate more in open source projects and work on the real-time scalable applications.

Yusuf Aytaş is a software engineer. He completed his B.S. and M.S. in computer science from Bilkent University, Ankara. He has worked in both early stage startups and multinational companies. He is proficient in agile methodologies, continuous delivery, and software development best practices.

About the Reviewers

Steve Fosdal has been writing software for over 10 years. He joined Slalom Consulting as a Solution Architect in late 2015.

His work has included building scalable, distributed applications for traffic prediction and real-time data integration using Akka, Scala, RabbitMQ, and Apache Spark.

He is also the primary contributor to camel-metrics, an open source Apache Camel component for route metrics.

I would like to thank my wife, Hilary, for her support and encouragement in everything that I do. Without her, I would not be able to be who I am today.

Van Thoai Nguyen has worked in the software industry for a decade in various domains. In 2012, he joined BuzzNumbers as one of the core senior software engineers, where he had opportunities to design, implement, and apply many cool technologies, tools, and frameworks. A RabbitMQ cluster was employed as the backbone of the real-time data processing platform, which includes various data collectors, data filtering, enrichment, and storage using a sharded cluster of MongoDB and SOLR. He is still maintaining the open source .NET RabbitMQ client library—Burrow.NET (`https://github.com/vanthoainguyen/Burrow.NET`)—which he built during the time he worked for BuzzNumbers. This library is still being used in many different applications in that company. He is interested in clean code and design, SOLID principle, and big data. You can find his blog at `http://thoai-nguyen.blogspot.com.au/`.

He is currently reviewing the book *Learning RabbitMQ* by Packt Publishing.

Jorge Puente-Sarrín is from Peru and is a software developer at RebelMouse. Prior to RebelMouse, he worked at Red Científica Peruana (RCP) and El Comercio, where he lead the adoption and integration of MongoDB into the company's IT infrastructure. He is a passionate developer focused on building distributed systems solutions using asynchronous programming with Python and .NET. Also, he has been contributing toward the translation of documentation projects and online courses into Spanish. He is a proud member of *Masters of MongoDB*, a group of persons promoting MongoDB around the world. He has also technically reviewed *RabbitMQ Cookbook*, by Packt Publishing.

Ken Taylor has worked in software development and technology for over 15 years. During the course of his career, he has worked as a systems analyst on multiple software projects for several industries as well as U.S. government agencies. He has successfully used RabbitMQ for messaging on multiple projects. He previously reviewed the books *RabbitMQ Cookbook* and *RabbitMQ Essentials* by Packt Publishing. He is a member and speaker of the 757 Ruby user's group and the Hampton Roads .NET user's group (HRNUG). He holds an A.S. in computer science from Paul D. Camp Community College and was awarded a U.S. Patent for a real estate financial software product. He is currently working at Outsite Networks Inc. in Norfolk, Virginia. He lives in Virginia Beach with his lovely wife, Lucia, and his two sons, Kaide and Wyatt.

I would like to thank my family for being a constant support in all of my endeavors.

Héctor Veiga Ortiz is a Software Engineer specializing in real-time data integration. Recently, he has focused his work on different cloud technologies (AWS, Heroku, OpenShift, etc.) to develop scalable, resilient and high-performing applications able to handle high-volume real-time data in diverse protocols and formats. Additionally, he has a strong foundation in messaging systems knowledge, such as RabbitMQ and AMQP. Lately, he has been focusing his work on the Akka, Apache Spark and Apache Flink. Also, Héctor has a master's degree in Telecommunication Engineering from the Universidad Politécnica de Madrid and a master's degree in Information Technology and Management from the Illinois Institute of Technology.

Héctor currently works at HERE as part of Global Data Integrations and is actively developing scalable applications to consume and preprocess data from several different sources. HERE heavily utilizes RabbitMQ to address their messaging requirements. In the past, Héctor worked at Xaptum Technologies, a company dedicated to M2M technologies.

Héctor has also worked on reviewing *RabbitMQ Cookbook*, *Learning RabbitMQ*, and *RabbitMQ Essentials* all by Packt Publishing.

I would like to thank Laura for her support. She keeps inspiring me and supporting me with everything I do. Without her, this would not have been possible.

www.PacktPub.com

Support files, eBooks, discount offers, and more

For support files and downloads related to your book, please visit www.PacktPub.com.

Did you know that Packt offers eBook versions of every book published, with PDF and ePub files available? You can upgrade to the eBook version at www.PacktPub.com and as a print book customer, you are entitled to a discount on the eBook copy. Get in touch with us at service@packtpub.com for more details.

At www.PacktPub.com, you can also read a collection of free technical articles, sign up for a range of free newsletters and receive exclusive discounts and offers on Packt books and eBooks.

https://www2.packtpub.com/books/subscription/packtlib

Do you need instant solutions to your IT questions? PacktLib is Packt's online digital book library. Here, you can search, access, and read Packt's entire library of books.

Why subscribe?

- Fully searchable across every book published by Packt
- Copy and paste, print, and bookmark content
- On demand and accessible via a web browser

Free access for Packt account holders

If you have an account with Packt at www.PacktPub.com, you can use this to access PacktLib today and view 9 entirely free books. Simply use your login credentials for immediate access.

Table of Contents

Preface

RabbitMQ is an open source messaging broker. It's often referred to as a message-oriented middleware that implements the Advanced Message Queuing Protocol (AMQP). Fundamentally, RabbitMQ provides a common platform for sending and receiving messages, where it guarantee the safety of messages until they are received. By playing an intermediary role between message consumers and producers, AMQP makes it easy to decouple applications.

Out of the box, RabbitMQ provides support for many messaging patterns. RabbitMQ guarantees data delivery, provides non-blocking operations, and sends push notifications. Moreover, it provides infrastructure for publish/subscribe, asynchronous processing, and work queues.

RabbitMQ provides a variety of features, including the tuning of application performance, clustering, flexible routing, federation, and so on. If you need specific features, RabbitMQ has several plugins that cater different needs. The RabbitMQ plugins extend its features in different ways, and you can also write your own plugin.

Through this book, we aim to give you a deep understanding of RabbitMQ and its use cases by providing multiple opportunities to learn about the message-oriented middleware, messaging architecture, messaging patterns, and solutions to real-life scenarios using RabbitMQ.

What this book covers

This book covers many aspects of software development with RabbitMQ. It provides thorough understanding of messaging, RabbitMQ, message-oriented software development, and so on.

Chapter 1, Getting Started, introduces you to message queues, message brokers, AMQP, and RabbitMQ.

Chapter 2, Configuring RabbitMQ, covers the configuration opportunities in RabbitMQ in detail.

Chapter 3, Architecture and Messaging, goes over RabbitMQ components—Producer, Message Broker, Consumer and the Message. This chapter provides learning opportunities for interoperability, heterogeneous integration, scalability, and so on.

Chapter 4, Clustering and High Availability, provides opportunities to tune RabbitMQ for high availability, federation, and much more.

Chapter 5, Plugins and Plugin Development, highlights several important features of RabbitMQ and gives an insight into creating your own plugin.

Chapter 6, Managing Your RabbitMQ Server, covers in detail the management of RabbitMQ using the command-line tools, management plugin, and rest API.

Chapter 7, Monitoring, discusses the methods to monitor RabbitMQ instances through a command line, management plugin, and well-known monitoring software.

Chapter 8, Security in RabbitMQ, covers the details about potential security vulnerabilities and securing RabbitMQ.

Chapter 9, Java RabbitMQ Client Programming, talks about developing RabbitMQ client using the Java platform.

Chapter 10, Ruby Client Programming, talks about developing a RabbitMQ client using Ruby.

Chapter 11, Python Client Programming, demonstrates developing a RabbitMQ client using Python.

What you need for this book

For this book, you need an understanding of software development—how to write functions, classes, and debugging skills. Moreover, you also need hands-on experience in developing applications.

Who this book is for

If you are an intermediate-level RabbitMQ developer and want to achieve professional-level expertise in the subject, this book is for you. You'll also need to have a decent understanding of message queuing.

Conventions

In this book, you will find a number of text styles that distinguish between different kinds of information. Here are some examples of these styles and an explanation of their meaning.

Code words in text, database table names, folder names, filenames, file extensions, pathnames, dummy URLs, user input, and Twitter handles are shown as follows: "We have the `dpkg` dependency management tool for installing RabbitMQ."

A block of code is set as follows:

```
tcp {
  upstream cluster {
    # simple round-robin
    server 192.168.1.1:5672;
    server 192.168.1.2:5672;
    check interval=3000 rise=2 fall=5 timeout=1000;
  }
  server {
    listen 5672;
    proxy_pass cluster;
  }
}
```

Any command-line input or output is written as follows:

```
mastering-rabbitmq1$ rabbitmqctl cluster_status
Cluster status of node rabbit@mastering-rabbitmq1 ...
[{nodes,[{disc,[rabbit@mastering-rabbitmq1]}]},
 {running_nodes,[rabbit@mastering-rabbitmq1]},
 {partitions,[]}]
...done.
```

New terms and **important words** are shown in bold. Words that you see on the screen, for example, in menus or dialog boxes, appear in the text like this: "In Windows, we should use the environment variables of the **System Properties** for modifying the environment variables of RabbitMQ."

[Warnings or important notes appear in a box like this.]

[Tips and tricks appear like this.]

Reader feedback

Feedback from our readers is always welcome. Let us know what you think about this book—what you liked or disliked. Reader feedback is important for us as it helps us develop titles that you will really get the most out of.

To send us general feedback, simply e-mail `feedback@packtpub.com`, and mention the book's title in the subject of your message.

If there is a topic that you have expertise in and you are interested in either writing or contributing to a book, see our author guide at `www.packtpub.com/authors`.

Customer support

Now that you are the proud owner of a Packt book, we have a number of things to help you to get the most from your purchase.

Downloading the example code

You can download the example code files from your account at `http://www.packtpub.com` for all the Packt Publishing books you have purchased. If you purchased this book elsewhere, you can visit `http://www.packtpub.com/support` and register to have the files e-mailed directly to you.

Errata

Although we have taken every care to ensure the accuracy of our content, mistakes do happen. If you find a mistake in one of our books—maybe a mistake in the text or the code—we would be grateful if you could report this to us. By doing so, you can save other readers from frustration and help us improve subsequent versions of this book. If you find any errata, please report them by visiting `http://www.packtpub.com/submit-errata`, selecting your book, clicking on the **Errata Submission Form** link, and entering the details of your errata. Once your errata are verified, your submission will be accepted and the errata will be uploaded to our website or added to any list of existing errata under the Errata section of that title.

To view the previously submitted errata, go to `https://www.packtpub.com/books/content/support` and enter the name of the book in the search field. The required information will appear under the **Errata** section.

Piracy

Piracy of copyrighted material on the Internet is an ongoing problem across all media. At Packt, we take the protection of our copyright and licenses very seriously. If you come across any illegal copies of our works in any form on the Internet, please provide us with the location address or website name immediately so that we can pursue a remedy.

Please contact us at `copyright@packtpub.com` with a link to the suspected pirated material.

We appreciate your help in protecting our authors and our ability to bring you valuable content.

Questions

If you have a problem with any aspect of this book, you can contact us at `questions@packtpub.com`, and we will do our best to address the problem.

1
Getting Started

Scalability is one of the major problems of our time, and messaging is an integral part of the solution. It finally comes down to the message broker software to manage and control messaging between applications, processes, and threads. Message brokers can help to solve scalability issues and architectural issues, such as coupling.

RabbitMQ is one of the most powerful open source message broker software that is widely used in the tech companies such as Mozilla, VMware, Google, AT&T, and so on. RabbitMQ is a highly configurable messaging platform developed and supported by a knowledgeable and committed community.

Before diving into the details and technologies behind the RabbitMQ, let's introduce you to the topics that we will cover in this chapter:

- A brief introduction to message brokers and the message queue
- An introduction to advanced message queue protocol
- Getting started with RabbitMQ
- Installing RabbitMQ
- Starting RabbitMQ
- Summary

Message brokers and message queue

Recently, software systems evolved dramatically. Applications have to communicate with other applications, these applications can be internal and external to the application itself. For the same application, we may have different type of clients, such as browsers, mobile clients, and so on. Hence, we absolutely need a communication layer between internal applications and between applications and clients. We need to deliver different messages to different applications or clients. Delivering messages can be a bottleneck if the communication layer isn't scalable. Pursuing scalable systems for communication layer leads us to **Message Brokers** and **Message Queues**. Let's now discuss what Message Brokers and Message Queues are.

Message brokers

A Message Broker is an architectural pattern that can receive messages from multiple destinations, determine the correct destination, and route the message along the correct route, as stated in the book *Enterprise Integration Patterns* by Hohpe and Woolf. Message brokers enable systems to deal with messaging and routing by mediating communication among components. Once applications implement a message broker pattern, it decreases the coupling between application components.

Message Brokers are centralized, in the architectural sense, to control and manage all messages. Therefore, all of the incoming and outgoing messages are sent through Message Brokers, which analyze and deliver the messages to their correct destination. This procedural step can be understood with the following diagram:

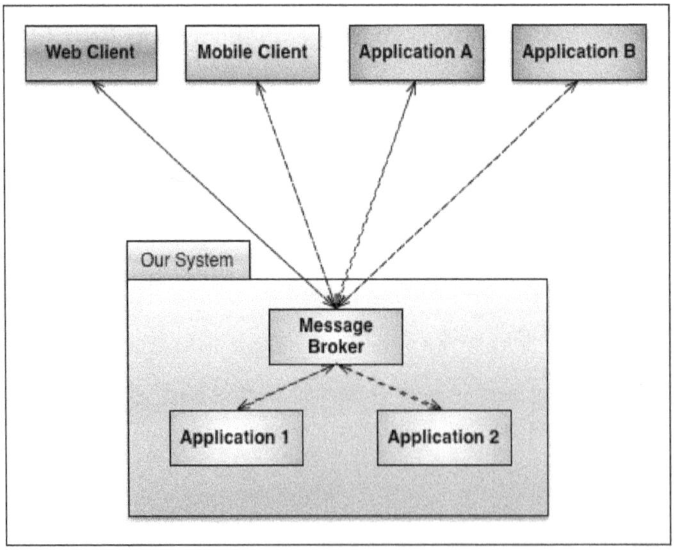

Message Broker

Message Brokers address the following concerns in the communication layer:

- Transforming messages to alternative formats
- Routing messages to destinations
- Supporting different types of patterns to send messages
- Receiving and responding to events
- Performing message aggregation
- Persisting the message states
- Ensuring the receiving and sending of message
- Decoupling the destination software systems

Many tasks of the Message Broker need a Message Queue for exchanging or passing data to the destination. The next section covers Message Queues. We will talk about the mechanism behind Message Brokers in *Chapter 3, Architecture and Messaging*.

Message Queues

A Message Queue is, briefly, a queue for messaging. Queue is the basic data structure behind the functioning of a Message Queue. Message Queue operations are similar to Queue data structure operations, such as the enqueue and dequeu operations. An enqueue operation leads to adding an element to the back of the queue. A dequeue operation leads to the deletion of an element from the front of the queue.

Message Queues provide concurrent and asynchronous operations to scale applications. In a message queue, messages wait up until a message is retrieved by an application. Let's take a look at the following diagram:

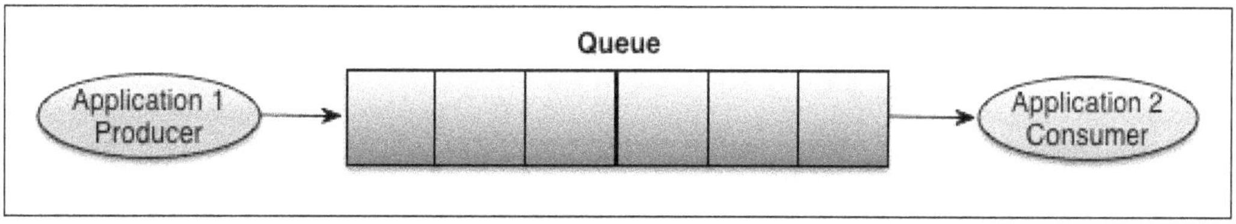

Message Queue

Different types of standards and protocols define the Message Queuing specifications. Some protocols are open to everyone; however, some protocols are closed. Let's come back to our topic. RabbitMQ uses **Advanced Message Queuing Protocol** (**AMQP**) that determines the policies of the Message Queues. The next topic will cover detailed information on AMQP. *Chapter 3, Architecture and Messaging,* covers the detailed explanation of Message Queues.

An introduction to the advanced message queue protocol

John O'Hara from J. P. Morgan started AMQP in 2003. He put incredible amount of work into it. Then, J. P. Morgan approached other firms to establish an organization for creating open standards in messaging. According to AMOP's official website (`http://www.amqp.org`), AMQP is an open standard for passing messages between applications or organizations. So, AMQP just defines the messaging properties, queue properties, how messages are routed between applications and clients, how Message Brokers ensure that the message is received or sent, and other concerns such as reliability and security.

According to the AMQP website (`http://www.amqp.org`), AMQP has lots of capabilities to accomplish goals:

- Security
- Reliability
- Interoperability
- Standard
- Open standard

Interoperability and reliability are very important for today's software engineering problems. The power of AMQP comes from its features like interoperability, reliability and so forth. Especially, with interoperability, we can use different types of technologies in sender and receiver. The main problem for most of the Internet giants is scalability. Scalability has direct relationship to reliability. *Chapter 3, Architecture and Messaging,* covers the details and specifications of AMQP.

An overview of RabbitMQ

As mentioned earlier, RabbitMQ is an open source Message Broker software that tries to solve messaging problems by implementing the AMQP. RabbitMQ is licensed with **Mozilla Public License**. RabbitMQ became part of **GoPivotal** in May 2013 and the community has helped in the development of RabbitMQ. Since then, the community has been trying to improve RabbitMQ.

RabbitMQ logo

As we explained, RabbitMQ solves messaging problems; however, you may ask yourself what kind of messaging problems RabbitMQ solves. In summary, RabbitMQ has the following functionalities to solve messaging problems:

- Ensures that messages are sent and received
- Routes the messages to the correct destinations
- Saves the state of the messages
- Supports multiple transportation protocols (AMQP, MQTT, STOMP, HTTP)
- Supports clustering
- Highly scalable, reliable and available
- Extendible with plugins
- Supports clients for almost any of language
- Large community support also provides commercial support

Now that we have discussed the concepts behind the RabbitMQ, let's try to install and use RabbitMQ on your computer.

Installation of RabbitMQ

Installation of RabbitMQ is not distinctly different from other software in different operation systems. Unix-based operating systems can build RabbitMQ from source code and Microsoft Windows can run the standard MSI installers. RabbitMQ installation files can be found in the download webpage of the RabbitMQ website, as shown in the following image:

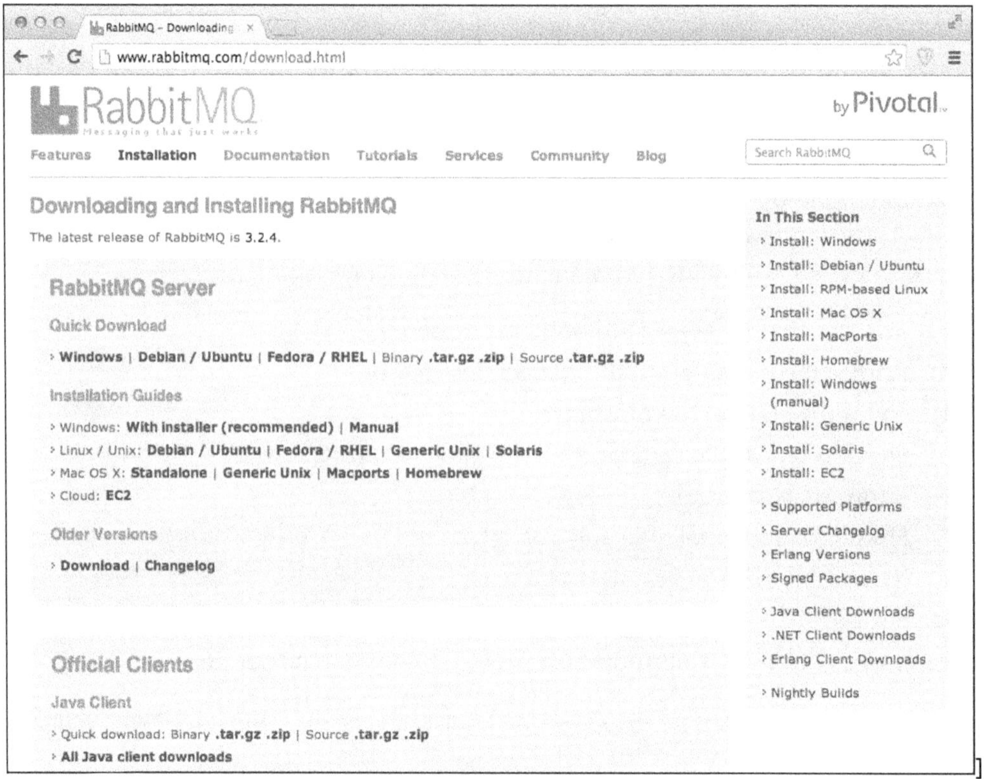

Download Webpage of RabbitMQ

The only prerequisite for the RabbitMQ installation is the **Erlang** runtime environment because RabbitMQ runs on the **Erlang VM**. Therefore, we have to install Erlang before installing the RabbitMQ. Erlang can be downloaded from the Erlang download webpage, as shown in the following image, and installation instructions will be covered in the topics that follow:

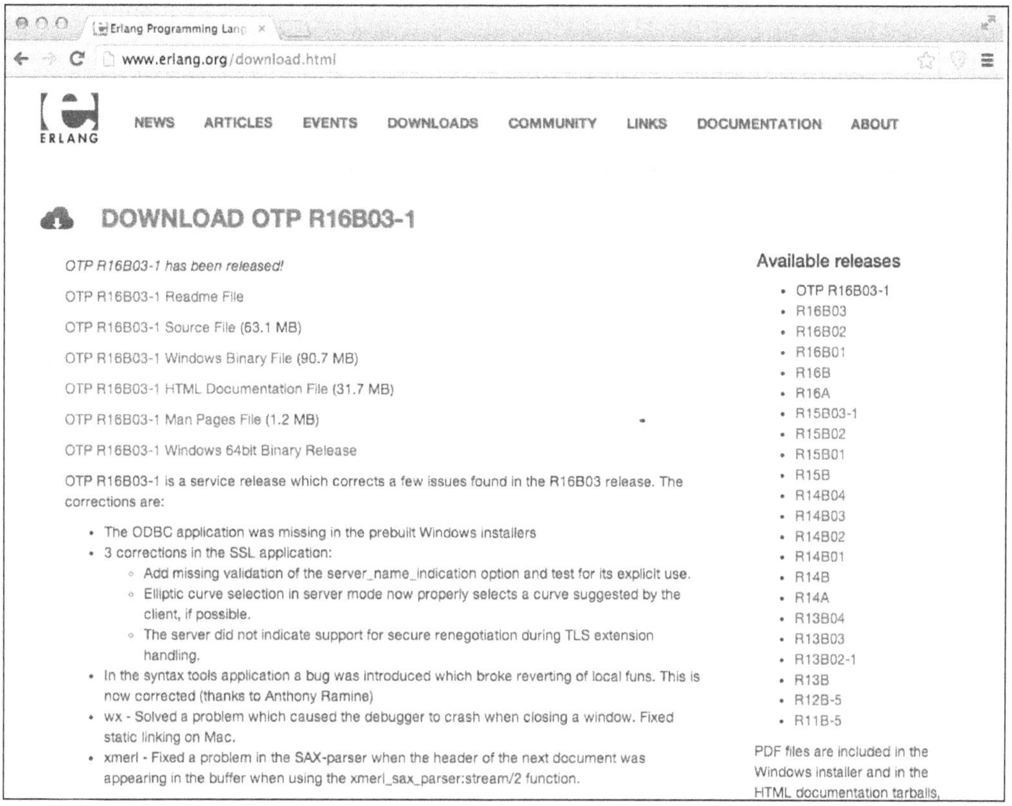

Download Webpage of Erlang

Now, let's to talk about the installation of both RabbitMQ and Erlang on Windows, Mac OS X, Ubuntu, Fedora, and Amazon Web Services.

Windows

RabbitMQ runs on both 32-bit and 64-bit machines from the same package. Erlang is installed either as 32-bit and 64-bit. So, RabbitMQ can be easily installed on the Windows operating system. Let's install these stuff for running RabbitMQ.

Firstly, we should install the Erlang runtime environment on Windows. Erlang has Windows installers for 32-bit and 64-bit as shown in the previous image. We can easily download the related binary file to our computer and install Erlang using it:

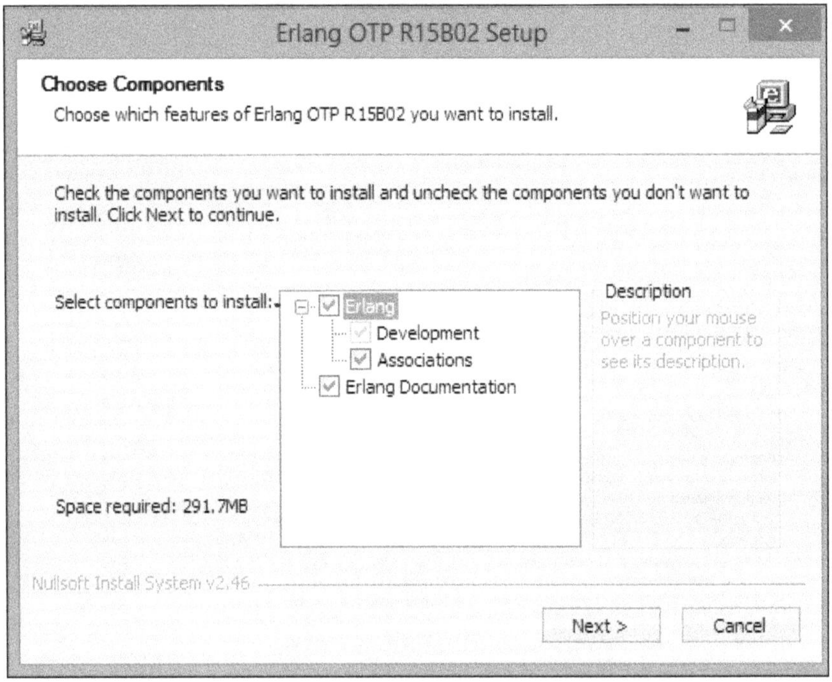

Installation of Erlang in Windows

After installing the Erlang runtime environment, we've completed the requirements of RabbitMQ installation. The next step is to download and install the RabbitMQ binary file with related the Windows version:

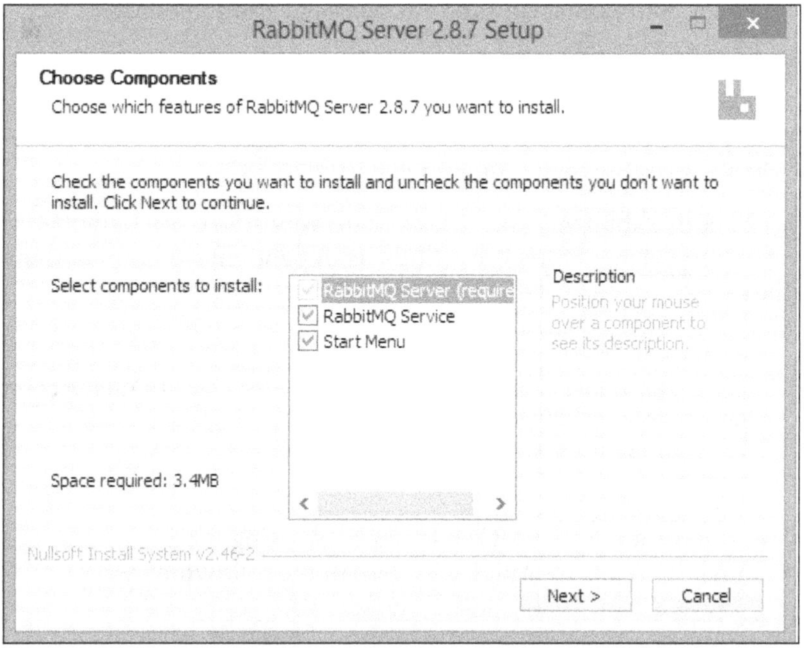

Installation of Erlang in Windows

We can find the related Windows installer for RabbitMQ with the help of RabbitMQ download webpage as shown in the screenshot showing the download webpage of RabbitMQ. Then, we just need to click and install the RabbitMQ on our Windows computer. Besides installing using the installer, we can install using the Windows binary file that is served within the RabbitMQ download webpage. The following instructions will be enough for installing RabbitMQ without the installer:

- Download the binary file for RabbitMQ Windows binary files
- Extract the downloaded RabbitMQ zip file to our local folder

It is possible to install the RabbitMQ on your Windows computer in both ways. Note that you may add the RabbitMQ binaries directory to the windows system path in the system/environment variable settings.

Mac OS X

As we specified, we only have one requirement to install RabbitMQ on our computers. In Mac OS X, we have package managers and we have the opportunity to compile from the source for both Erlang runtime environment and RabbitMQ.

Firstly, both Erlang and RabbitMQ can be easily installed on Mac OS X using package managers. Although we have lots of package managers on Mac OS X, **Homebrew** and **MacPort** are the ones that are mostly used in Mac OS X. So, we'll talk about the installation using Homebrew and MacPorts.

Homebrew has both RabbitMQ and Erlang on its repository. As RabbitMQ has a dependency with Erlang, Homebrew finds its dependent software and installs them together.

> Homebrew is just another package manager for Mac OS X. Homebrew is quite easy to install on Mac OS X and has lots of packages. So, you would find your application in its repository. Check it out at `http://brew.sh`

So, we just need to install RabbitMQ in Homebrew using the `brew install rabbitmq` command on our terminal as shown in the following image:

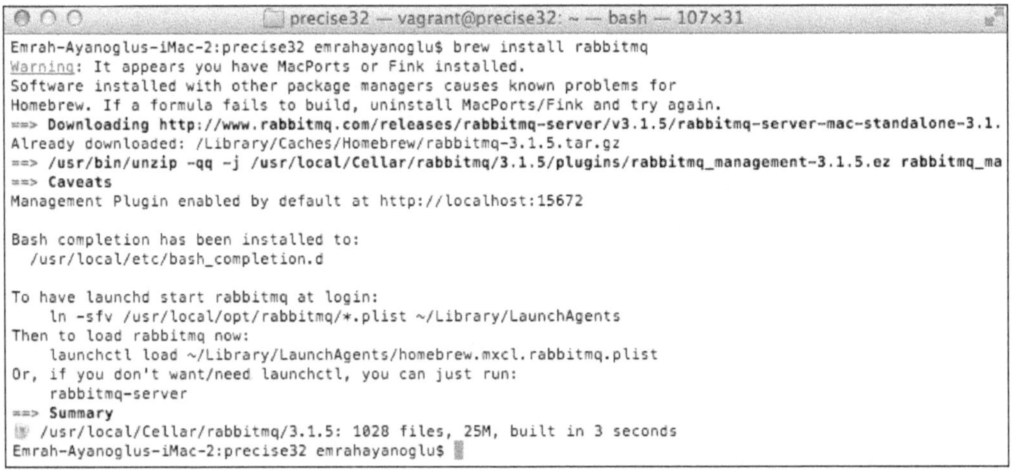

precise32 — vagrant@precise32: ~ — bash — 107×31
Emrah-Ayanoglus-iMac-2:precise32 emrahayanoglu$ brew install rabbitmq
Warning: It appears you have MacPorts or Fink installed.
Software installed with other package managers causes known problems for
Homebrew. If a formula fails to build, uninstall MacPorts/Fink and try again.
==> Downloading http://www.rabbitmq.com/releases/rabbitmq-server/v3.1.5/rabbitmq-server-mac-standalone-3.1.
Already downloaded: /Library/Caches/Homebrew/rabbitmq-3.1.5.tar.gz
==> /usr/bin/unzip -qq -j /usr/local/Cellar/rabbitmq/3.1.5/plugins/rabbitmq_management-3.1.5.ez rabbitmq_ma
==> Caveats
Management Plugin enabled by default at http://localhost:15672
Bash completion has been installed to:
/usr/local/etc/bash_completion.d
To have launchd start rabbitmq at login:
ln -sfv /usr/local/opt/rabbitmq/*.plist ~/Library/LaunchAgents
Then to load rabbitmq now:
launchctl load ~/Library/LaunchAgents/homebrew.mxcl.rabbitmq.plist
Or, if you don't want/need launchctl, you can just run:
rabbitmq-server
==> Summary
/usr/local/Cellar/rabbitmq/3.1.5: 1028 files, 25M, built in 3 seconds
Emrah-Ayanoglus-iMac-2:precise32 emrahayanoglu$

Homebrew Installation of RabbitMQ

MacPorts has the similar method of operation with Homebrew. MacPorts also installs the software with its dependencies. Therefore, we just need to install RabbitMQ in MacPorts using the `port install rabbitmq-server` command on our terminal, as shown in the following image:

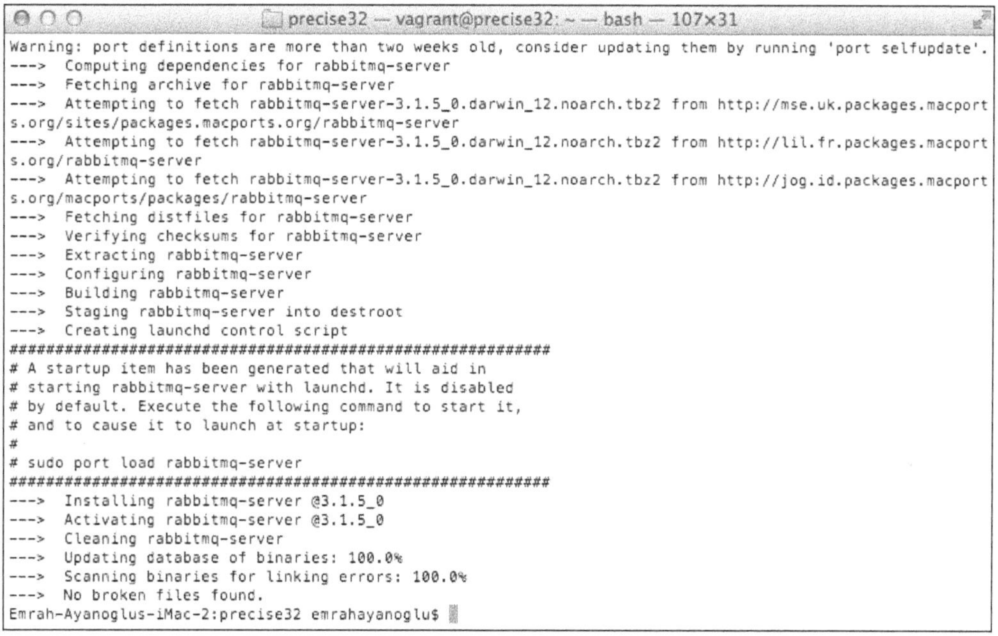

precise32 — vagrant@precise32: ~ — bash — 107×31
Warning: port definitions are more than two weeks old, consider updating them by running 'port selfupdate'.
---> Computing dependencies for rabbitmq-server
---> Fetching archive for rabbitmq-server
---> Attempting to fetch rabbitmq-server-3.1.5_0.darwin_12.noarch.tbz2 from http://mse.uk.packages.macports.org/sites/packages.macports.org/rabbitmq-server
---> Attempting to fetch rabbitmq-server-3.1.5_0.darwin_12.noarch.tbz2 from http://lil.fr.packages.macports.org/rabbitmq-server
---> Attempting to fetch rabbitmq-server-3.1.5_0.darwin_12.noarch.tbz2 from http://jog.id.packages.macports.org/macports/packages/rabbitmq-server
---> Fetching distfiles for rabbitmq-server
---> Verifying checksums for rabbitmq-server
---> Extracting rabbitmq-server
---> Configuring rabbitmq-server
---> Building rabbitmq-server
---> Staging rabbitmq-server into destroot
---> Creating launchd control script
###
A startup item has been generated that will aid in
starting rabbitmq-server with launchd. It is disabled
by default. Execute the following command to start it,
and to cause it to launch at startup:
#
sudo port load rabbitmq-server
###
---> Installing rabbitmq-server @3.1.5_0
---> Activating rabbitmq-server @3.1.5_0
---> Cleaning rabbitmq-server
---> Updating database of binaries: 100.0%
---> Scanning binaries for linking errors: 100.0%
---> No broken files found.
Emrah-Ayanoglus-iMac-2:precise32 emrahayanoglu$

MacPorts Installation of RabbitMQ

Another way to install RabbitMQ and Erlang is by compiling the source codes in Mac OS X. Before compiling, we need the following to compile Erlang source code:

- The GNU make
- The GNU C compiler
- Perl 5

After downloading and unzipping the source codes of the Erlang, we just need common commands on the `Erlang` folder for compiling from source code, as follows:

```
./configure
make
make install
```

Finally, we just need to download and unzip the RabbitMQ binary files.

Ubuntu

Ubuntu is just another Linux distribution based on Debian. Similar instructions as the ones we discussed for the installation on the Mac OS X would be applied for Ubuntu.

Ubuntu has a package manager called **Advanced Packaging Tool** (**apt-get**) and has a Debian package manager called **dpkg**. So, we are able to install RabbitMQ and Erlang runtime environment using `apt-get`. Moreover, similar to Mac OS X, we can compile from source codes of Erlang.

Firstly, as we said in the previous paragraph, we can install RabbitMQ using `apt-get` and `dpkg`. Before installing RabbitMQ, we should add the RabbitMQ repository to the APT repository using the following line (add the following line to `/etc/apt/sources.list`):

```
deb http://www.rabbitmq.com/debian/ testing main
```

Now, we are ready to install RabbitMQ and its dependency Erlang runtime environment, as shown in the following image:

```
sudo apt-get install rabbitmq-server
```

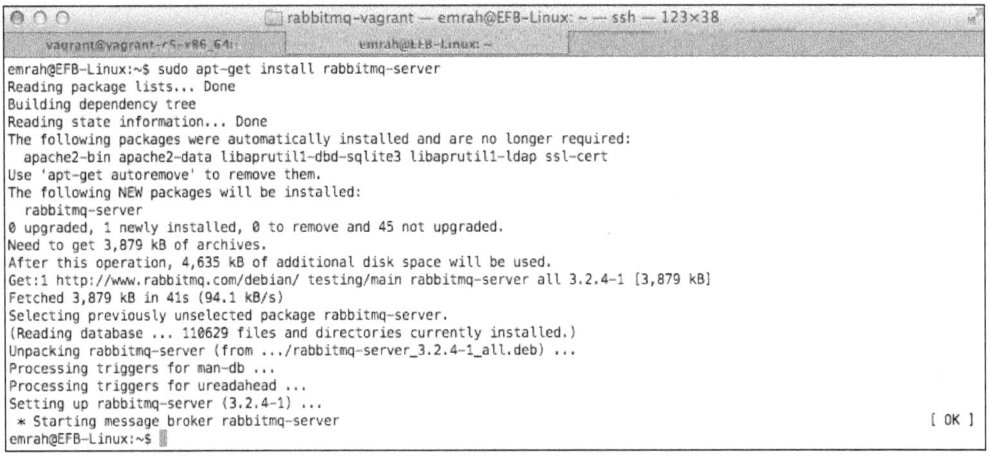

rabbitmq-vagrant — emrah@EFB-Linux: ~ — ssh — 123×38
emrah@EFB-Linux:~$ sudo apt-get install rabbitmq-server
Reading package lists... Done
Building dependency tree
Reading state information... Done
The following packages were automatically installed and are no longer required:
apache2-bin apache2-data libaprutil1-dbd-sqlite3 libaprutil1-ldap ssl-cert
Use 'apt-get autoremove' to remove them.
The following NEW packages will be installed:
rabbitmq-server
0 upgraded, 1 newly installed, 0 to remove and 45 not upgraded.
Need to get 3,879 kB of archives.
After this operation, 4,635 kB of additional disk space will be used.
Get:1 http://www.rabbitmq.com/debian/ testing/main rabbitmq-server all 3.2.4-1 [3,879 kB]
Fetched 3,879 kB in 41s (94.1 kB/s)
Selecting previously unselected package rabbitmq-server.
(Reading database ... 110629 files and directories currently installed.)
Unpacking rabbitmq-server (from .../rabbitmq-server_3.2.4-1_all.deb) ...
Processing triggers for man-db ...
Processing triggers for ureadahead ...
Setting up rabbitmq-server (3.2.4-1) ...
* Starting message broker rabbitmq-server [OK]
emrah@EFB-Linux:~$

Ubuntu Installation of RabbitMQ

Also, we have the `dpkg` dependency management tool for installing RabbitMQ. RabbitMQ has packages for `dpkg` in its download webpage. We can download it from its website then run the following command:

```
dpkg -i rabbitmq-server.deb
```

Secondly, we have another option, which was explained in the Mac OS X topic. That is, compiling from source codes. We just need to compile the downloaded Erlang source code, and we are ready to run the downloaded binary files of RabbitMQ. You can look at the details in the Mac OS X section.

Fedora

Fedora is yet another Linux distribution based on Red Hat. Installation instructions of Fedora are similar to Ubuntu's installation. Fedora has package managers called **rpm** and **yum**.

Firstly, we are able to install RabbitMQ with its dependency Erlang using package managers. Before using `yum`, we should run the following command to add RabbitMQ repository:

```
wget -O /etc/yum.repos.d/epel-erlang.repo
http://repos.fedorapeople.org/repos/peter/erlang/epel-erlang.repo
```

Then, we can install RabbitMQ using the following command as shown in the following screenshot:

```
sudo yum install rabbitmq-server
```

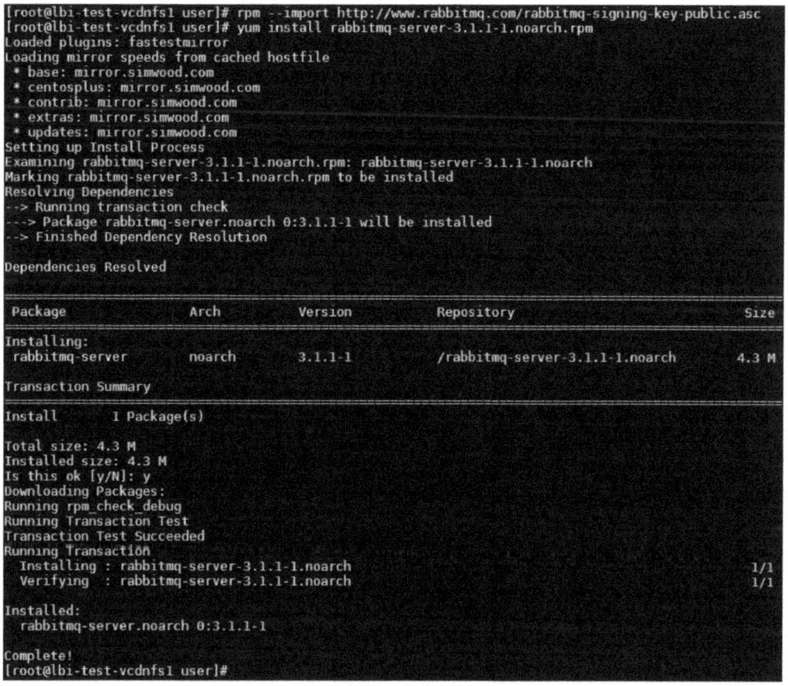

Fedora Installation of RabbitMQ

Fedora has another package manager, which comes from Red Hat, called rpm. As RabbitMQ publishes package as `rpm`, we can easily install using the `rpm` package manager. After downloading the `rpm` package from RabbitMQ webpage, we can install RabbitMQ with Erlang, using following command:

```
rpm -ivh rabbitmq-server.rpm
```

Secondly, we have another option, which was explained in the Mac OS X topic. That is compiling from source code. We just need to compile the downloaded Erlang source code and we are ready to run the downloaded binary files of RabbitMQ. You can look at the details in the Mac OS X section.

Amazon elastic compute cloud (EC2)

Amazon Web Services (**AWS**) is cloud-computing platform offered from Amazon. AWS has lots of features for developers such as **autoscaling**. Besides RabbitMQ, AWS also offers their own messaging service called **Simple Queue Service**: however, RabbitMQ has lots of advantages over Amazon SQS. For instance, RabbitMQ has an extendable plugin system, whereas SQS capabilities are short. RabbitMQ implements a standard approach to message acknowledgement and consumption, whereas SQS has its own standards.

Anyway, we can easily install our RabbitMQ to the AWS EC2 instance and can save images of the RabbitMQ installed operating system. In AWS EC2, we choose one of the operating systems from the list or any other instance that we used earlier.

> Amazon EC2 would be a good choice for your servers. In a scalable architecture, we need clusters of Message Brokers, databases, caches, and so on. EC2 gives you great API, and it's really quite easy to create clusters using EC2. We will talk about the clusters of RabbitMQ using EC2 in *Chapter 4, Clustering and High Availability*.

As we explained, we are able to install RabbitMQ on Windows, Linux, and Mac OS X, we just need to follow the instructions for AWS EC2 that are explained well in the preceding sections. Let's take a look at the following screenshot:

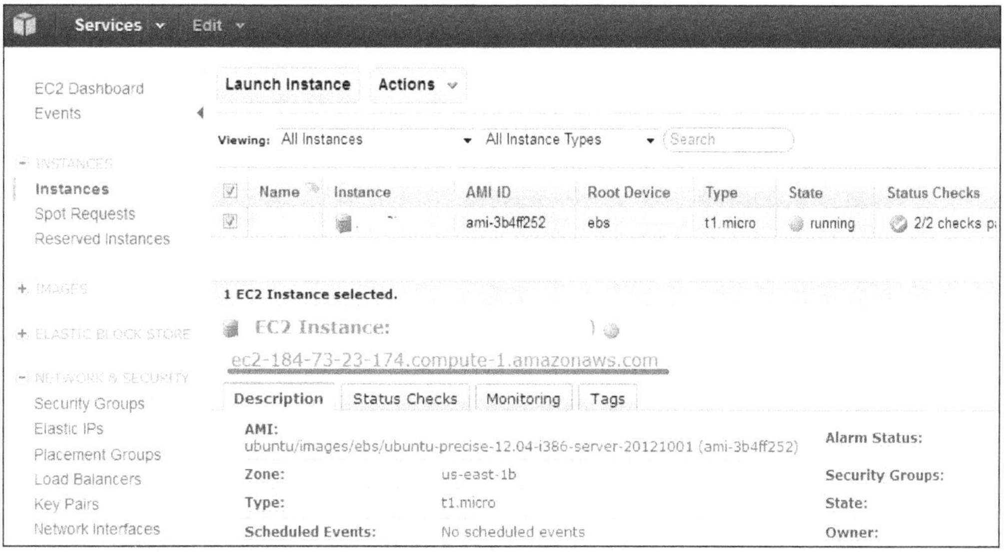

Amazon EC2

Starting RabbitMQ

As we can see, the installation part of RabbitMQ is quite easy and starting RabbitMQ is similar to its installation. Some package managers in Linux, Mac OS X, and Windows installer add configuration parameters to operation system's configuration for automatic startup. In such a case, we don't need to run the RabbitMQ command manually; however, if we install RabbitMQ manually, we need to run the RabbitMQ commands manually.

Starting RabbitMQ on Windows

If we use the Windows installer of RabbitMQ, the installer already makes configurations for starting automatically. Therefore, we don't need to run RabbitMQ manually; however, whenever we'd like to control the status of the server, we just need to run following command on the `sbin` folder of RabbitMQ:

```
rabbitmqctl status
```

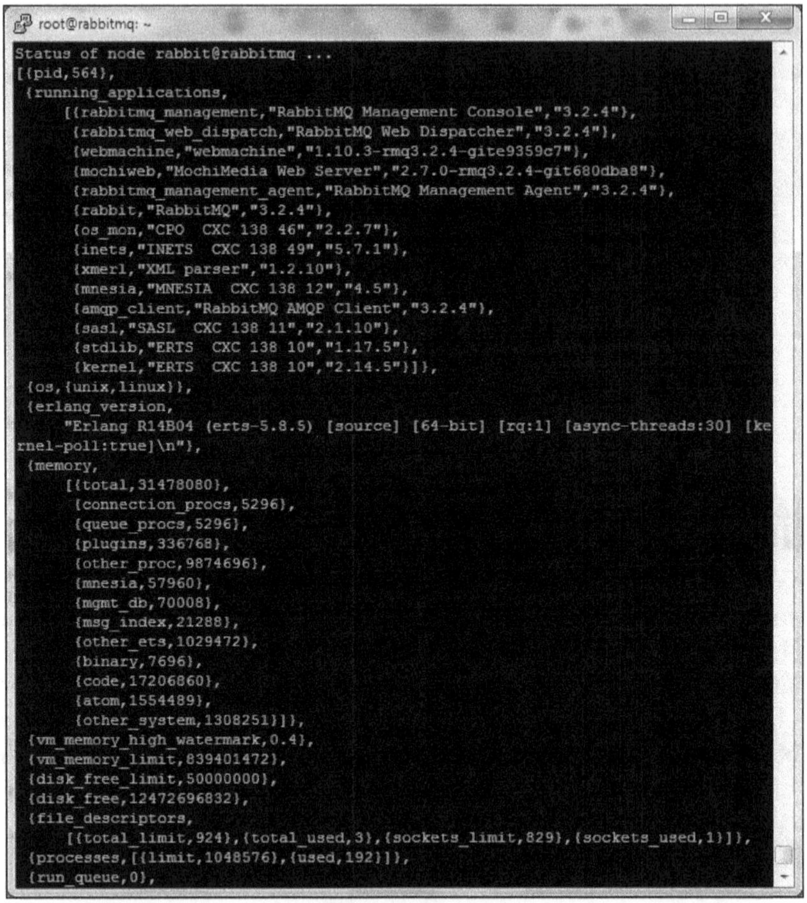

Status of RabbitMQ in Windows

You may have installed RabbitMQ manually on your Windows and you might wonder how you can run the RabbitMQ server. You should run the following command to start RabbitMQ (you have to run this command with an administrative user). Moreover, you can install the RabbitMQ server as a Windows service:

```
rabbitmq-server
```

Other OSes (Linux, Mac OS X)

There's isn't much difference in running RabbitMQ on Windows and other operating systems. If we have RabbitMQ already installed using package managers, such as `apt-get`, `yum`, and so on, we don't need to run the RabbitMQ manually because RabbitMQ has already started automatically. So, we'd like to check the status of the RabbitMQ using the following command:

```
sudo rabbitmqctl status
```

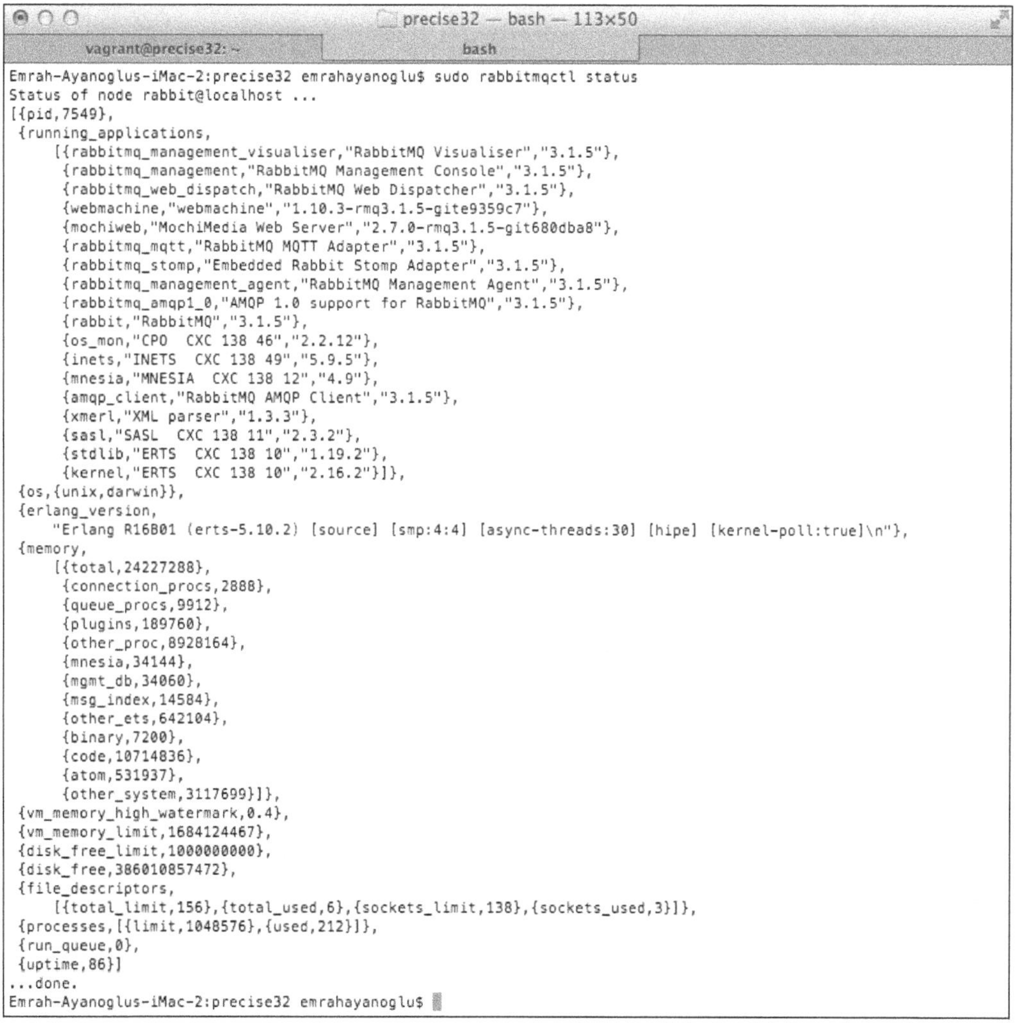

```
precise32 — bash — 113×50
vagrant@precise32: ~    bash
Emrah-Ayanoglus-iMac-2:precise32 emrahayanoglu$ sudo rabbitmqctl status
Status of node rabbit@localhost ...
[{pid,7549},
 {running_applications,
     [{rabbitmq_management_visualiser,"RabbitMQ Visualiser","3.1.5"},
      {rabbitmq_management,"RabbitMQ Management Console","3.1.5"},
      {rabbitmq_web_dispatch,"RabbitMQ Web Dispatcher","3.1.5"},
      {webmachine,"webmachine","1.10.3-rmq3.1.5-gite9359c7"},
      {mochiweb,"MochiMedia Web Server","2.7.0-rmq3.1.5-git680dba8"},
      {rabbitmq_mqtt,"RabbitMQ MQTT Adapter","3.1.5"},
      {rabbitmq_stomp,"Embedded Rabbit Stomp Adapter","3.1.5"},
      {rabbitmq_management_agent,"RabbitMQ Management Agent","3.1.5"},
      {rabbitmq_amqp1_0,"AMQP 1.0 support for RabbitMQ","3.1.5"},
      {rabbit,"RabbitMQ","3.1.5"},
      {os_mon,"CPO  CXC 138 46","2.2.12"},
      {inets,"INETS  CXC 138 49","5.9.5"},
      {mnesia,"MNESIA  CXC 138 12","4.9"},
      {amqp_client,"RabbitMQ AMQP Client","3.1.5"},
      {xmerl,"XML parser","1.3.3"},
      {sasl,"SASL  CXC 138 11","2.3.2"},
      {stdlib,"ERTS  CXC 138 10","1.19.2"},
      {kernel,"ERTS  CXC 138 10","2.16.2"}]},
 {os,{unix,darwin}},
 {erlang_version,
     "Erlang R16B01 (erts-5.10.2) [source] [smp:4:4] [async-threads:30] [hipe] [kernel-poll:true]\n"},
 {memory,
     [{total,24227288},
      {connection_procs,2888},
      {queue_procs,9912},
      {plugins,189760},
      {other_proc,8928164},
      {mnesia,34144},
      {mgmt_db,34060},
      {msg_index,14584},
      {other_ets,642104},
      {binary,7200},
      {code,10714836},
      {atom,531937},
      {other_system,3117699}]},
 {vm_memory_high_watermark,0.4},
 {vm_memory_limit,1684124467},
 {disk_free_limit,1000000000},
 {disk_free,386010857472},
 {file_descriptors,
     [{total_limit,156},{total_used,6},{sockets_limit,138},{sockets_used,3}]},
 {processes,[{limit,1048576},{used,212}]},
 {run_queue,0},
 {uptime,86}]
...done.
Emrah-Ayanoglus-iMac-2:precise32 emrahayanoglu$
```

Status of RabbitMQ in Unix

After controlling the RabbitMQ status, if we get a message that says that RabbitMQ isn't running, then we should run RabbitMQ using the following command on the `sbin` folder of RabbitMQ installation folder:

```
rabbitmq-server
```

precise32 — vagrant@precise32: ~ — beam.smp — 82×31
Emrah-Ayanoglus-iMac-2:precise32 emrahayanoglu$ sudo rabbitmq-server
RabbitMQ 3.1.5. Copyright (C) 2007-2013 GoPivotal, Inc.
Licensed under the MPL. See http://www.rabbitmq.com/
##
########## Logs: /usr/local/var/log/rabbitmq/rabbit@localhost.log
/usr/local/var/log/rabbitmq/rabbit@localhost-sasl.log
##########
Starting broker... completed with 10 plugins.

Starting of RabbitMQ in Unix

Summary

We finally finished our first chapter, which introduces messaging concepts along with brief details about Message Queues and Message Brokers, and the protocol called AMQP that defines the functionalities of a Message Queue. Finally, our chapter introduces the RabbitMQ, providing information about its installation on different types of operating systems and how we can run the RabbitMQ server. Now, we are ready to jump into the details of RabbitMQ, starting from its configuration in the next chapter.

2
Configuring RabbitMQ

Configuration is one of the crucial parts for administrating RabbitMQ. With an excellent configuration, RabbitMQ can send and receive messages effectively between applications, processes, and threads.

There are three ways to configure RabbitMQ. The first way is to use the RabbitMQ environment variables that lie on the environment variables of the operating system, the second way is through the configuration file provided by RabbitMQ, and the last way is to use runtime parameters. This configuration diversity gives full control of RabbitMQ on server side and operating system side.

This chapter covers the overall idea behind the configuration of RabbitMQ and three ways to configure it. So, we'll cover the following points:

- The overall configuration of RabbitMQ
- The RabbitMQ environment variables
- The configuration file
- The runtime parameters

Overall configuration of RabbitMQ

RabbitMQ's configuration is quite important to ensure the performance, high availability, and scalability within the installed operating system. In order to utilize RabbitMQ, we have three configuration ways:

- **Environment variables**: These are specified in the networking parameters and file locations
- **Configuration file**: This expresses the server component settings for permissions, limits, plugins, and clusters

- **Runtime parameters**: These define the cluster settings that would change at run time

Before diving into each configuration type, we should check whether the configuration file exists or not. In Unix-based systems, you can find the configuration file in the following folder:

```
/etc/rabbitmq/rabbitmq.config
```

In Windows, you can find the configuration file in the following folder:

```
C:\Program Files (x86)\RabbitMQ_Server\etc\rabbitmq.config
```

After checking the configuration file, we are now ready to talk about configuration types one by one.

The RabbitMQ environment variables

RabbitMQ environment variables is one of the configuration ways of RabbitMQ. Every operating system has its own set of environment variables for each user. Although operating systems has the ability to have environment variables, the way in which environment variables are changed is slightly different across operating systems.

In Unix-based operating systems, we can change the environment variables rather easily using the `rabbitmq-env.conf` file. In the environment configuration file, we can add the environment parameters as follows:

```
CONFIG_FILE=/etc/rabbitmq/testfile
```

After changing the `rabbitmq-env.conf` file, we have to restart the RabbitMQ server to reload the environment variables.

In Windows, we should use the environment variables of the **System Properties** for modifying the environment variables of RabbitMQ. We can access to the **Environment Variables** by navigating to **Settings** | **Control Panel** | **System Properties** | **Advanced** | **Environment Variables**, where we use pipes to show transitions. You can see this in the following screenshot:

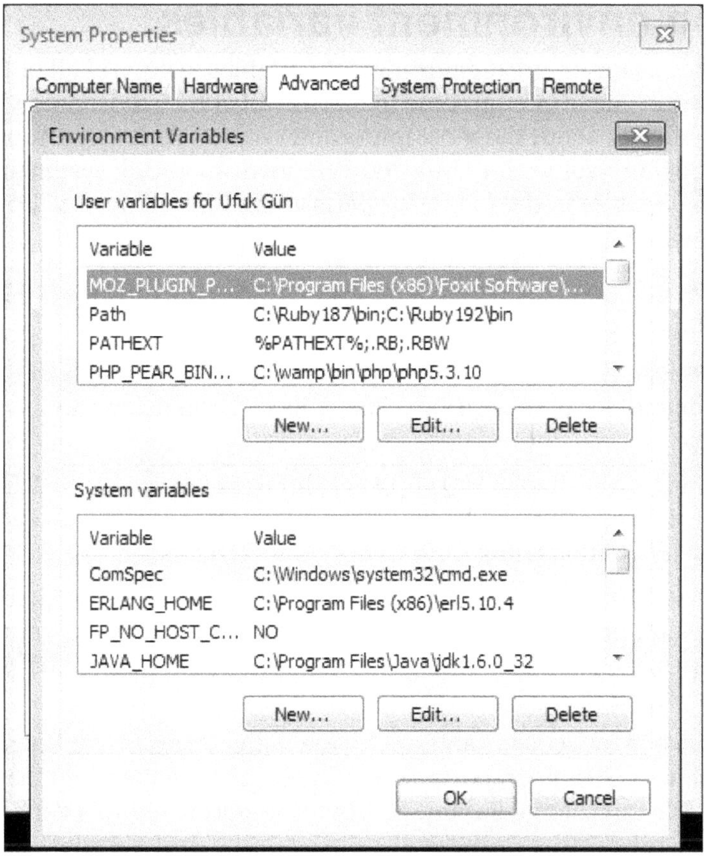

Windows Environment Parameters

We are now accessing the environment variables of different operating systems. Although, RabbitMQ gives us lots of different environment variables, we will cover the most important ones.

Common environment variables

RabbitMQ gives us lots of great environment variables to control all of the parts of its engine. We don't have enough time to discuss all of the environment variables; however, we will talk about the most important ones. Furthermore, some variables have different default values for Unix and Windows operating systems; we'll consider these variables in the following parts of the topic. Anyway, let's dive into the important variables:

- `RABBITMQ_BASE`: This variable basically locates the directory of RabbitMQ. This directory has the database and log files.
- `RABBITMQ_CONFIG_FILE`: Although the `configuration` file of RabbitMQ has a default location, you can change its location using this environment variable.
- `RABBITMQ_LOGS`: RabbitMQ supports different levels of logs. Whenever RabbitMQ is creating a `log` file, it has a default location; however, you can change its location using this environment variable.
- `RABBITMQ_NODE_IP_ADDRESS`: RabbitMQ binds to all network interfaces as a default property. As RabbitMQ gives us a full control over network interfaces, we can easily change its binding network using this variable, such as `127.0.0.1`.
- `RABBITMQ_NODE_PORT`: RabbitMQ has a default port, `5672`; however, we have sometimes collision on ports, so we should change the ports that RabbitMQ binds. We can change RabbitMQ's binding port using this variable.
- `RABBITMQ_PLUGINS_DIR`: RabbitMQ has many very useful plugins that will be enabled through RabbitMQ. RabbitMQ has a default location for these Erlang coded plugins; however, you can change its location.

The RabbitMQ Environment Variables:

Name	Default Value	Description
`RABBITMQ_BASE`	`*(default)`	This is the directory in which RabbitMQ server's database and log files are located.
`RABBITMQ_CONFIG_FILE`	`*`	This is the name of `configuration` file. The name doesn't consist of the extension "`.config`".

Name	Default Value	Description
RABBITMQ_CONSOLE_LOG		This variable can have one of the two values: "new" or "reuse". These variables are used to decide the console log file whether create a new log file or reuse the old log file. If these variables are not set, the console output will not be saved.
RABBITMQ_LOGS	*	This is the directory of the RabbitMQ log file.
RABBITMQ_LOG_BASE	*	This is the base directory that holds the log files. If RABBITMQ_LOGS or RABBITMQ_SASL_LOGS is set, then this variable has no effect on configuration.
RABBITMQ_MNESIA_BASE	*	This expresses the base location of the Mnesia databases files. If RABBITMQ_MNESIA_DIR is set, then this variable has no effect on configuration.
RABBITMQ_MNESIA_DIR	*	This variable specifies the location of Mnesia database files.
RABBITMQ_NODE_IP_ADDRESS	The empty string means that this binds to all network interfaces.	This is the binding address. You should change this attribute when you'd like to bind to a single network interface.
RABBITMQ_NODENAME	On Unix: rabbit@hostname On Windows: rabbit@%COMPUTERNAME%	This is the node name of RabbitMQ server. This should be unique per Erlang node and machine combination.

Name	Default Value	Description
RABBITMQ_NODE_PORT	5672	This is the binding port of RabbitMQ server.
RABBITMQ_PLUGINS_DIR	*	The location where plugins of RabbitMQ server are located.
RABBITMQ_SASL_LOGS	*	This is the location of RabbitMQ server's **System Application Support Libraries**' log files.
RABBITMQ_SERVICENAME	On Windows Service: RabbitMQ On Unix: rabbitmq-server	This variable specifies the service name that is installed on the service system of operating system.
RABBITMQ_SERVER_START_ARGS	None	Erlang parameters are used for the erl command when invoking the RabbitMQ server. This variable will not override RABBITMQ_SERVER_ERL_ARGS.

Cells marked with* will be explained in the Unix and Windows section

Unix-specific default location

The following table describes the Unix-specific default locations of the given environment variables. Most of the locations are related to the installed location.

Default locations of environment variables for Unix:

Name	Location
RABBITMQ_BASE	This variable is not used for Unix
RABBITMQ_CONFIG_FILE	${install_prefix}/etc/rabbitmq/rabbitmq
RABBITMQ_LOGS	$RABBITMQ_LOG_BASE/$RABBITMQ_NODENAME.log
RABBITMQ_LOG_BASE	${install_prefix}/var/log/rabbitmq
RABBITMQ_MNESIA_BASE	${install_prefix}/var/lib/rabbitmq/mnesia

Name	Location
RABBITMQ_MNESIA_DIR	$RABBITMQ_MNESIA_BASE/$RABBITMQ_NODENAME
RABBITMQ_PLUGINS_DIR	$RABBITMQ_HOME/plugins
RABBITMQ_SASL_LOGS	$RABBITMQ_LOG_BASE/$RABBITMQ_NODENAME-sasl.log

Windows-specific default location

In contrast to Unix, Windows default values of the RabbitMQ environment variables are related to the other environment variables of RabbitMQ. The following table shows the Windows default locations.

Default locations of environment variables for Windows:

Name	Location
RABBITMQ_BASE	%APPDATA%\RabbitMQ
RABBITMQ_CONFIG_FILE	%RABBITMQ_BASE%\rabbitmq
RABBITMQ_LOGS	%RABBITMQ_LOG_BASE%\%RABBITMQ_NODENAME%.log
RABBITMQ_LOG_BASE	%RABBITMQ_LOG_BASE%\log
RABBITMQ_MNESIA_BASE	%RABBITMQ_BASE%\db
RABBITMQ_MNESIA_DIR	%RABBITMQ_MNESIA_BASE%\%RABBITMQ_NODENAME%
RABBITMQ_PLUGINS_DIR	%RABBITMQ_BASE%\plugins
RABBITMQ_SASL_LOGS	%RABBITMQ_LOG_BASE%\%RABBITMQ_NODENAME%-sasl.log

RabbitMQ environment variables are highly dependent on operating system environment variables. As an example, Computer Name in Unix and Hostname in Windows set the environment variable `RABBITMQ_SERVICENAME` and `RABBITMQ_NODENAME`. Anyway, the following table describes the dependent environment variables:

Name	Default Value	Description
COMPUTERNAME	Unix: env hostname	This is the name of current machine for Windows machines.

Name	Default Value	Description
`ERLANG_SERVICE_MANAGER_PATH`	Windows Service: `%ERLANG_HOME%\erts-x.x.x\bin`	This is the location where the Erlang service wrapper script is located.
`HOSTNAME`	Windows: `localhost`	This is the name of the current location for Unix machines.

The configuration file

The RabbitMQ environment variables mostly gives the control of location of files and directories, whereas the RabbitMQ configuration file gives the control of the engine, such as authentication, performance, memory limit, disc limit, exchanges, queues, bindings, and so on. The `configuration` file is by default located in `/etc/rabbitmq/rabbitmq.config` for Unix-based computers and `$RABBITMQ_SERVER\etc\rabbitmq.config` for Windows-based computers, as discussed in the previous sections.

RabbitMQ has many configuration variables; however, we will discuss the most important ones here:

- `auth_mechanisms`: This is used to supports different types of authentication mechanisms. You can change the different type of authentication mechanism using this variable.
- `default_user`: This is used as a default user to access the RabbitMQ server using the RabbitMQ client. The `default_user` variable simply defines the username of the default user.
- `default_pass`: This is similar to the `default_user` variable, as it simply defines the default user's password.
- `default_permission`: This is similar to `default_user` and `default_pass`. This variable describes the permissions of the default user.
- `disk_free_limit`: This is used to controls the disk size for storing the messages into the disk. This variable defines the free disk size to give an alert to the RabbitMQ server administrator.

- heartbeat: This is a configuration variable that indicates the time interval between beats. A beat is a packet sent from the broker to the client, and back so that the broker can understand whether a client is still connected or not and to keep a line open where some network equipment may cut it due to inactivity. In other protocols, such as the old and time-tested **Internet Relay Chat (IRC)** , this trick was also known as **ping-pong**.
- hipe_compile: As a default property, RabbitMQ is compiled with the default Erlang compiler; however, we can compile with the high performance Erlang compiler. RabbitMQ server is compiled at startup. Hipe compiling results in later start; however, Hipe Compile gives 20%-40% performance gain on message broker operations. With hipe_compile variable, we can control whether RabbitMQ will be compiled through high performance Erlang compiler or not.
- log_levels: We have logs to control and trace the application for each of the software applications. Logs have different levels to show log messages according to its log level, that is, error, warning, and information. With this variable, you can decide on the log level.
- tcp_listeners: This is the same as the server applications, such as FTP server, Mail server, and so on. The RabbitMQ server binds on the IP and port of the operating system. You can change its binding port and IP with the tcp_listeners variable.
- ssl_listeners: Whenever clients listen to the server with SSL, the RabbitMQ server uses a different IP and port. The ssl_listeners variable just defines the IP and port of the SSL client connections.
- vm_memory_high_watermark: Free memory size is reasonably important for the RabbitMQ server. RabbitMQ alerts the memory problem with the given free memory ratio in vm_memory_high_watermark.

The following table describes the most of the important variables with given default values:

Configuration Variables:

Variable Name	Description
auth_mechanisms	This variable specifies the SASL authentication mechanisms. Default value: ['PLAIN', 'AMQPLAIN']

Variable Name	Description
auth_backends	This variable specifies the authentication databases to use in SASL. Other databases would be used with this plugin support. Default value: [rabbit_auth_backend_internal]
collect_statistics	This variable specifies the statistics collection mode. Default value: none Possible values: • none • coarse • fine
collect_statistics_interval	This variable specifies the statistics collection interval in miliseconds. Default value: 5000
default_pass	This variable specifies the default password for the RabbitMQ server to create a user in a scratched database. Default value: Guest
default_permission	This variable specifies the default permissions of the default user. Default value: [".*", ".*", ".*"]
default_user	This variable specifies the default username for the RabbitMQ server to create a user in a scratched database.
disk_free_limit	This variable specifies the disk's free space limit of the partition on which RabbitMQ has stored the data. If available disk space is lower than the disk free limit, then flow control is triggered. Moreover, the value should be related to the memory size. Default value: 50000000

Variable Name	Description
heartbeat	This variable specifies the heartbeat delay in seconds. Default value: 580 Possible values: 0 means heartbeats are disabled
hipe_compile	This variable specifies whether precompile parts of RabbitMQ with the high performance Erlang compiler or not. This variable directly affects the performance of the message rate. Hipe is supported only on Unix-based machines. Default value: False
log_levels	This variable specifies the granularity of logging. Default value: [{connection, info}] Possible values: • none • error • warning • info
msg_store_file_size_limit	This variable specifies the file size limit of storing each message. Default value: 16777216
tcp_listeners	This variable specifies the ports that listen for AMQP connections without SSL. This variable may contain integers like 5672 that describes only the port and dictionary structure that describes both the IP and the port, for example, {"127.0.0.1", 5672}. Default value: [5672]

Variable Name	Description

tcp_listen_options	This variable specifies the socket options.
	Default value:
	`[binary, {packet, raw},`
	`{reuseaddr, true},`
	`{backlog, 128},`
	`{nodelay, true},`
	`{exit_on_close, false}]`

server_properties	This variable specifies the key-value pairs that is to announce to clients on starting connection Default Value: `[]`
ssl_listeners	This variable specifies the ports that listen for AMQP connections with SSL. This variable may contain integers like `5672` that describes only the port and dictionary structure that describes both the ip and port, such as, `{"127.0.0.1", 5672}`. Default value: `[]`
ssl_options	This variable specifies the configuration for the SSL type. Default value: `[]`
reverse_dns_lookup	This variable specifies whether RabbitMQ performs a reverse DNS lookup on client connections or not. Default value: `False`
vm_memory_high_watermark	This variable specifies the memory threshold. Default value: `0.4`, that is, `4/10`

Runtime parameters

RabbitMQ provides environment variables and configuration variables to configure RabbitMQ when starting the RabbitMQ server. In addition to these configurations, RabbitMQ allows us to change parameters, which were set in the environment variables and configuration variables in the runtime using the runtime parameters.

We can use the command-line tool for managing the RabbitMQ broker for changing the runtime parameters, as shown in the following screenshot:

```
emrahayanoglu — bash — 85×73
Emrah-Ayanoglus-iMac-2:~ emrahayanoglu$ rabbitmqctl help
Error: could not recognise command
Usage:
rabbitmqctl [-n <node>] [-q] <command> [<command options>]

Options:
    -n node
    -q

Default node is "rabbit@server", where server is the local host. On a host
named "server.example.com", the node name of the RabbitMQ Erlang node will
usually be rabbit@server (unless RABBITMQ_NODENAME has been set to some
non-default value at broker startup time). The output of hostname -s is usually
the correct suffix to use after the "@" sign. See rabbitmq-server(1) for
details of configuring the RabbitMQ broker.

Quiet output mode is selected with the "-q" flag. Informational messages are
suppressed when quiet mode is in effect.

Commands:
    stop [<pid_file>]
    stop_app
    start_app
    wait <pid_file>
    reset
    force_reset
    rotate_logs <suffix>

    join_cluster <clusternode> [--ram]
    cluster_status
    change_cluster_node_type disc | ram
    forget_cluster_node [--offline]
    update_cluster_nodes clusternode
    sync_queue queue
    cancel_sync_queue queue

    add_user <username> <password>
    delete_user <username>
    change_password <username> <newpassword>
    clear_password <username>
    set_user_tags <username> <tag> ...
    list_users

    add_vhost <vhostpath>
    delete_vhost <vhostpath>
    list_vhosts [<vhostinfoitem> ...]
    set_permissions [-p <vhostpath>] <user> <conf> <write> <read>
    clear_permissions [-p <vhostpath>] <username>
    list_permissions [-p <vhostpath>]
    list_user_permissions <username>

    set_parameter [-p <vhostpath>] <component_name> <name> <value>
    clear_parameter [-p <vhostpath>] <component_name> <key>
    list_parameters [-p <vhostpath>]

    set_policy [-p <vhostpath>] <name> <pattern>  <definition> [<priority>]
    clear_policy [-p <vhostpath>] <name>
    list_policies [-p <vhostpath>]
```

Image 2: Command Line Tool for Managing a RabbitMQ broker

Parameter management

Parameter management is a way to configure RabbitMQ by setting the parameter values. We are able to change parameters using the `set_parameter` command of `rabbitmqctl`. Moreover, we can change the different types of components of RabbitMQ with the given `component_name` attribute. The following tables shows the parameters and description of the runtime parameters:

Parameter	Description
`set_parameter [-p vhostpath] {component_name} {name} {value}`	The `set_parameter` parameter performs the setting of the parameters of a given component at runtime.
`clear_parameter [-p vhostpath] {component_name} {key}`	The `clear_parameter` parameter removes all of the parameters of a given component at runtime.

Policy management

Policy management is configuration of the RabbitMQ policy values. RabbitMQ gives us an opportunity to change its policies for message queues in the runtime, and its policies are applicable for exchange and queues. You can set the new policies using "`set_policy`", whereas you can clear all the policies using "`clear_policy`".

Parameter	Description
`set_policy [-p vhostpath] [--priority priority] [--apply-to apply-to] {name} {pattern} {definition}`	The `set_policy` parameter performs a change of the behavior of the queues and exchanges by setting the given pattern and definition.
`clear_policy [-p vhostpath] {name}`	The `clear_policy` parameter removes all of the policies, which is given with the `name` parameter.

Memory management

Memory management is the configuration of RabbitMQ memory values. Memory management can be done through the RabbitMQ configuration file parameters as we saw in the previous sections. But, sometimes we have to change the memory threshold to a lower value for many client attractions. RabbitMQ gives an option to the change memory threshold using the "`set_vm_memory_high_watermark`" runtime parameter as shown in the following table:

Parameter	Description
`set_vm_memory_high_watermark {fraction}`	The `set_vm_memory_high_watermark` parameter changes the memory threshold fraction.

Summary

Configuration is an important part of administrating RabbitMQ. We can achieve a decent configuration over RabbitMQ using its different configuration types to different parts of RabbitMQ. We can control the file locations and network configuration using environment variables; express the server settings, authentication, permissions, and limits through configuration variables in `configuration` file of RabbitMQ; and finally, we can change these parameters at the runtime using runtime parameters.

Now, we have completed the configuration part of RabbitMQ. We are ready to dive into the technical and architectural structure of RabbitMQ in the next chapter.

3
Architecture and Messaging

RabbitMQ server simply solves messaging problems. But what is the meaning of messaging itself? Sometimes, the term messaging is confused with real-time messaging such as chat messages, SMS messages, and so on. These systems also have the messaging system in their subsystems; however, we are talking about a somewhat different issue.

By the dictionary definition, messaging is a short communication transmitted by words, signals, or other means from one person, station, or group to another. In computer engineering, definition of messaging seems like the dictionary definition. Messaging simply takes the messages from a producer and sends it to the consumers by the computer engineering definition. In messaging systems, we are using some architecture related to messaging and elements. Moreover, we have different functionalities of the elements in the messaging system. So, we will start our most important chapter with the messaging concepts and their roots, and then we'll dive into the details of AMQP that describes the RabbitMQ mechanism. The following list shows the structure of this chapter:

- Messaging and use cases of messaging
- Enterprise messaging
- Messaging related software architectures
- Messaging concepts
- Advanced message queuing protocol (AMQP)

Messaging and its use cases

As we discussed in the introduction of this chapter and in the first chapter, messaging is simply defined as communication between the message producer and the consumer of the message. **Message broker** is defined as a module that controls messaging flow. Controlling action isn't that simple, so message brokers needs lots of skills to accomplish this messaging functionality.

Before talking about the Message Broker functionalities, we need to know the problems that we have with the messaging. The problems change with respect to the domain of the software system; however, most of the problems are the same within the different types of messaging functionalities of the software systems. Let's list all of the common problems of messaging and how we solve these problems using message brokers.

Coupling of the software systems

Nowadays, coupling is generally referred to as the expression of dependency between two modules with respect to each other. Coupling, or more specifically tangling code, is bad because it makes it harder to maintain software. Any change on dependent component may result in changes, bugs, version upgrades, and so on. Tight coupling can be at the code level, and at the service/architecture level. For both code and architectural coupling, solutions exist. Code level coupling can be solved with dependency injection. Architectural coupling can be solved with message brokers. We need to create an abstraction between modules for messaging issue. The following screenshot shows us the coupled messaging modules that interact with the other modules without using Message Broker. Whenever you change one method of the module, you have to propagate this change to the other modules:

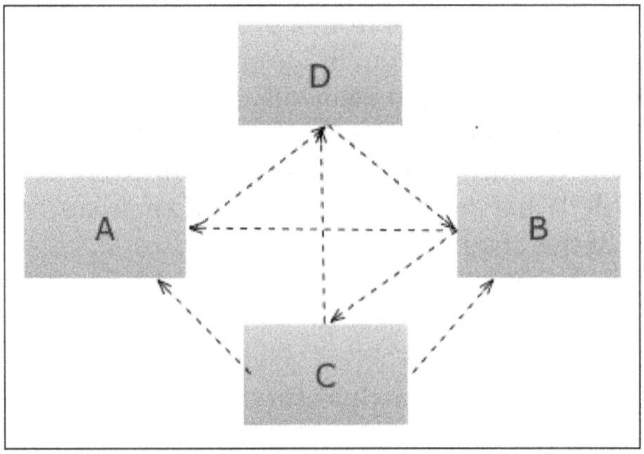

Coupling of messaging modules

Message Broker is an awesome solution for the well-known problem of high-coupled software systems that is communicated between modules. Message Broker creates an abstraction between modules, so that messaging functionality is controlled and managed by the Message Broker itself. Modules are not aware of the sending or receiving of messages; they only send their messages to the right receiver via Message Broker. Message Broker routes these messages to the right module and transforms them to the appropriate messaging format.

As a consequence, Message Broker is defined as messaging middleware that simply makes a transition of high-coupled messaging modules to the low-coupled messaging modules by creating the intermediate layer between modules. As shown in the following screenshot, modules are sending and receiving messages from each other without knowing the functionality of the intermediate layer, which sends the messages to **Message Broker** and routes them to the right module:

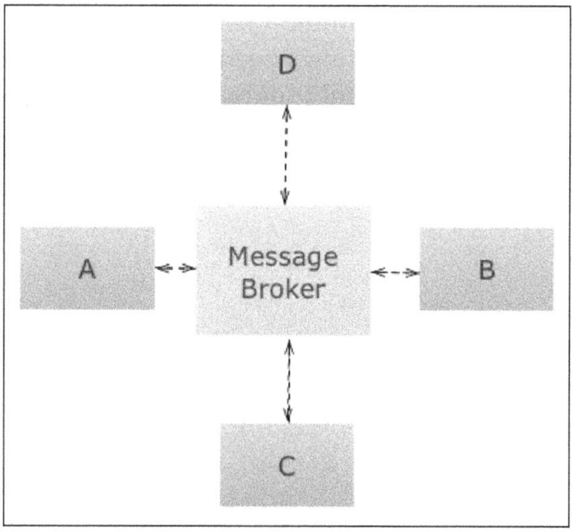

Decoupling of messaging modules using message broker

Heterogeneous integration

Nowadays, every software systems are using different types of technologies, such as Java Platform, .Net Framework, Mainframes, and so on. Also, the mobility and web gives us the opportunity to add new clients to our software systems. Therefore, we have to merge these technology stacks in software architecture and connect them to each other; for instance, connecting Java Platform to .Net Framework. Then, we have another problem in our software system: **Heterogeneous Integration**.

We have great web service solutions to guarantee the Heterogeneous Integration between technology stack and different types of clients; however, all of them have bottlenecks; so we have another great solution: Message Broker. Message Broker has a capability to send and receive messages without analyzing the details of the sender and receiver. Message Broker just sends the messages in a given format to the right module. The receiver module just needs to parse the message into their format. Thus, Message Broker just gives us the combination of different stacks into one architecture, and it gives us another solution for well-known software engineering problem: **Interoperability**. The following screenshot will show you how Message Broker communicates between different programming languages:

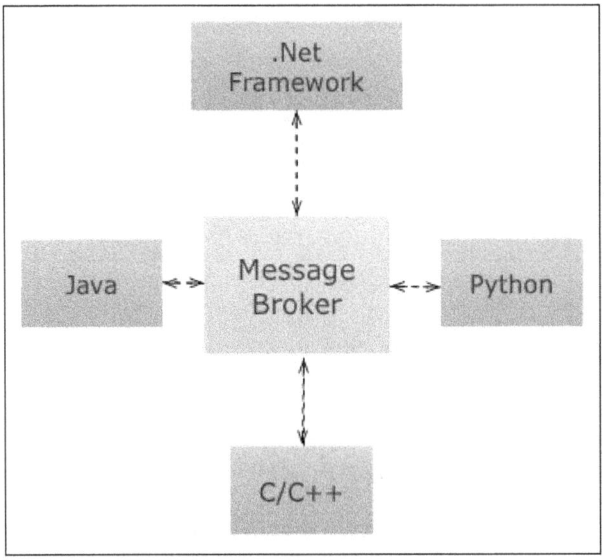

Heterogeneous Integration

As a consequence, not only Message Broker sends the messages to right receiver, but it also satisfies the Heterogeneous Integration to combine different types of technologies in our stack.

Addressing scalability

In dictionary definition, **Scalability** simply means that the software or network system can adapt to increased demands. Scalability is our modern software engineering problem; lots of academic papers have aimed to solve this well-known issue. Scalability is not an easy problem and doesn't have a single solution. Although scalability differs from financial software systems to real time web applications, they have a similar problem: communicating between modules or processes. Then it comes to Message Brokers, which solve this scalability problem.

We all know that Message Brokers can address part of the scalability problem, but we need to know how it addresses this. Message Brokers gives us these items in its muscle:

- It increases the overall throughput of the system, which reduces the response time of the software system
- Multiple message receiver and sender capability of Message Broker gives us an opportunity to cope with concurrent messages
- It allows us to create an Asynchronous System, so we have an opportunity to control messaging in our modules in an event-driven manner

Consequently, Message Broker eases well-known problem scalability by delivering high throughput, less response time, and highly concurrent system.

So, we showed the three different problems that we aim to solve using Message Brokers. Message Brokers not only solve these well-known problems, but also aim to solve all of the messaging problems between applications, modules, and processes. Note that messaging itself can become single point of failure. Scaling messaging is where brokers are expected to be effective.

Enterprise messaging

Enterprise applications such as customer relationship management applications, business intelligence applications, project management applications, human resource management applications, and so on have to integrate with other enterprise applications. You can see the overall diagram of communication between enterprise applications with the following screenshot. Moreover, we have to guarantee that all systems should be scalable. So, we have problems such that we talked before.

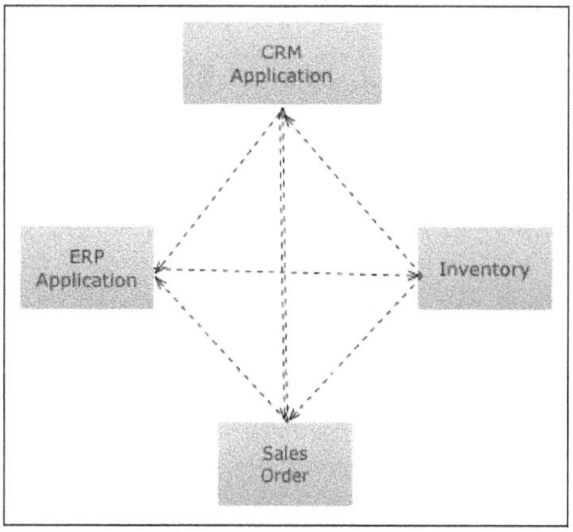

Sample financial application before message broker

Message Brokers mainly attempt to solve these kinds of problems in enterprise applications. In enterprise messaging, we have to guarantee that the message is sent and received, since each of the messages is very important for our system's robustness. Message Brokers have a functionality to store all messages permanently to satisfy this kind of requirement. You can see the overall diagram of the software architecture with the following screenshot:

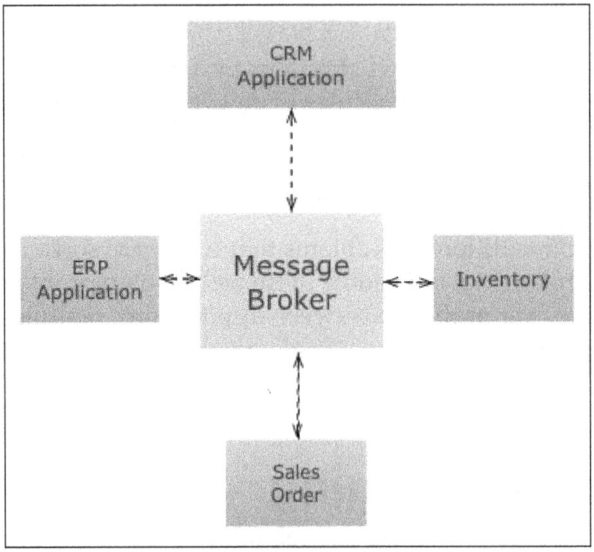

Sample financial application after message broker

As a result, enterprise messaging is not very different from the other types of messaging applications. Thus, we are able to solve enterprise-messaging problems using the Message Brokers.

Messaging-related software architectures

As we went through how problems are addressed with Messaging and Message Brokers, we are now ready to dive into the software architectures that are parallel with these solutions. Message Brokers create an abstraction between modules, applications, and processes. The abstraction, which is created with Message Brokers, is simply defined as **Message Oriented Middleware**. As we know, Message Brokers has a functionality to apply asynchronous patterns on the affected modules, applications, and processes. Then, it comes to **Event Driven Architecture** that is strictly related with the asynchronous system. After defining these related software architectures, let's dive into this amazing stuff.

Message oriented middleware – Architecture

Message Oriented Middleware is simply defined as a component that allows software components, which have been placed on the same or different network, to communicate with one another. In a Producer/Consumer pattern, producers send their message to different consumers with the help of Message Oriented Middleware, guaranteeing the message received as shown in the next screenshot. If we look at the definition deeply, Message Oriented Middleware tries to solve some software engineering problems such as interoperability, monitoring, enterprise integration of software systems, abstraction, reliability, security, and so on.

Interoperability is defined as systems and devices that can exchange data without knowing each other's functionality. In an interoperable system, we have heterogeneous software components rather than homogeneous components. Message Oriented Middleware integrates all of the heterogeneous components and interacts with them in a scalable way. As we talked before, we are now using different technologies in our technology stack. With the help of Message Oriented Middleware, we are able to interact with these different types of technologies.

Application performance management is widely described as monitoring and management of performance and availability of software applications. Nowadays, we need to monitor performance metrics closely in most software systems. Message Oriented Middleware can ease up monitoring and tuning performance. So, both consumers and producers are partly relieved from problems like monitoring, logging, and tuning.

Today, we are using different software systems from different companies. Especially in financial area, we have to use customer relationship management software, enterprise resource planning software, human resource management software, and so on. Also, every director needs to combine these software systems in a single software architecture. This is done by enterprise integration of software systems. This integration has many software architectures, and Message Oriented Middleware is one of the architectures that satisfy the enterprise integration of software systems.

As a consequence, Message Oriented Middleware that is provided with the Message Brokers is responsible to solve some software engineering problems such as interoperability, enterprise integration of software systems, and so on. We are able to use Message Oriented Middleware in our software system whenever using the Message Brokers. You can see an overview of message oriented architecture in the following screenshot:

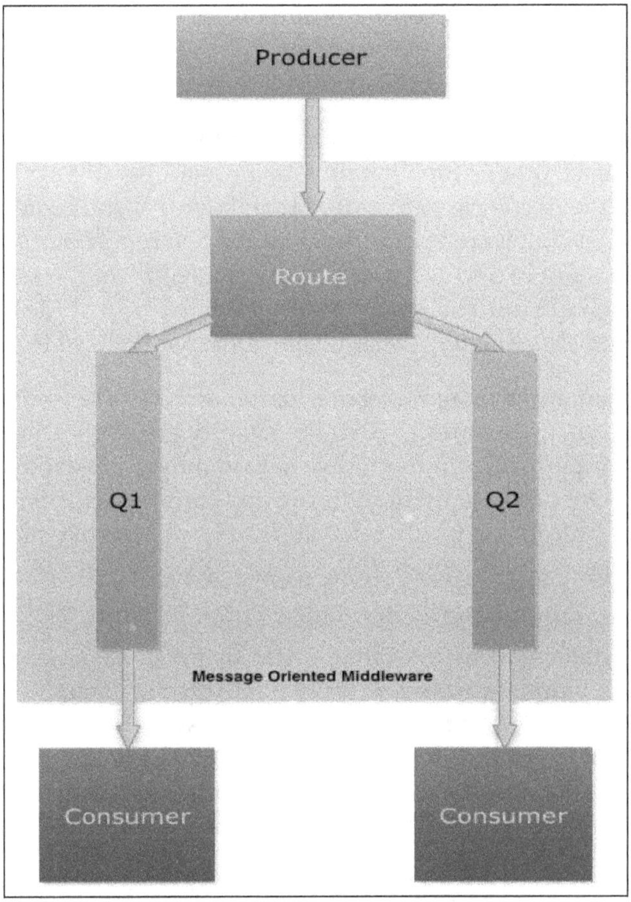

Message Oriented Middleware architecture

Event-driven architecture

In an asynchronous system, operations take place independent of other operations; therefore, operations can take place without waiting for others. Since Message Broker's support asynchronous operations, they can be easily used in an Event Driven Architecture. Before defining **Event Driven Architecture** (**EDA**), let's talk about the problems before EDA.

As we discussed in the messaging concept, producers send message to the consumers. Before EDA, consumers always await their incoming messages from producers. Then, one of the processes had to control the waiting process. With EDA, we have listeners, whose duty is to trigger the listener event whenever a message is received. Now we are ready to define the EDA. EDA is a push-based communication between producer and consumer, which gives reaction to the events.

The structure of EDA consists of four elements: event creator or producer, event consumer, event manager, and event. The following is an explanation of these elements:

- Event creator is just the source of event
- Event consumer is a listener of event that needs to know the event has occurred
- Event manager is a middleware between creator and consumer, which is the controller of the events and triggers the related event consumers
- Event is an action that is detected by a Event listener or consumer.

These structures could be seen in an example in the preceding screenshot. Note that arrows between components are events.

EDA solves lots of software engineering problems, for example, scalability, high availability, and so on. So, it is good to list the benefits of the EDA:

- EDA has a capacity to support large numbers of creators and consumers
- It responses to information in near real time
- It prevents the blocking or waiting in the consumer phase
- It shapes the architecture as an extremely loosely-coupled architecture

Consequently, Message Brokers support asynchronous software system. Then, we are able to use Event Driven Architecture in our software system to gather great solutions to well-known problems such as scalability. You can see an overview of event driven architecture in the following screenshot:

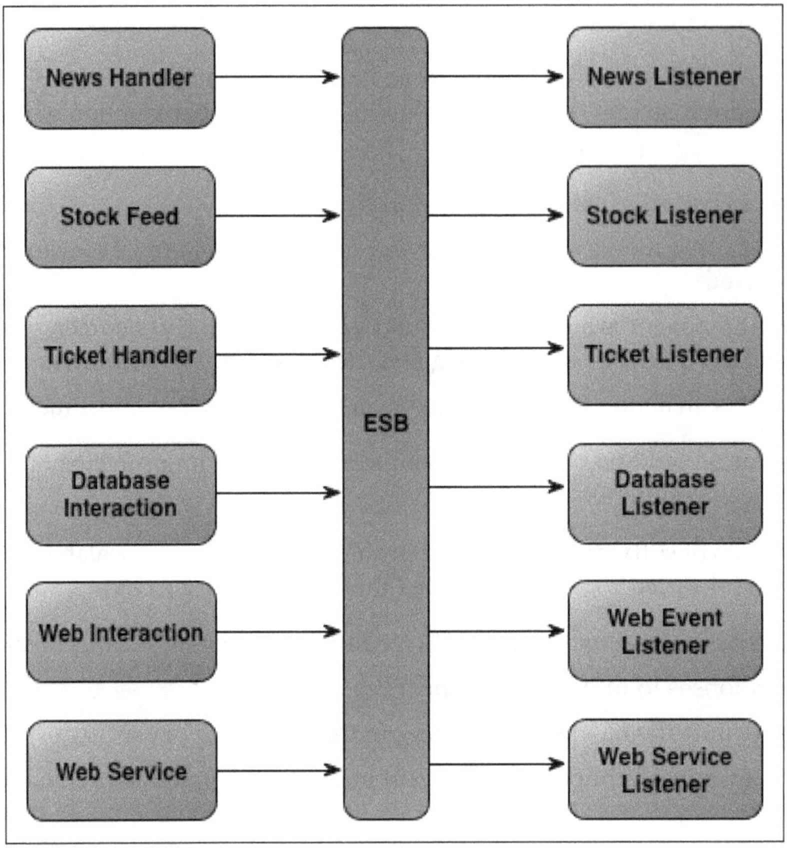

Image 7: Event Driven Architecture

Messaging concepts

So far, we talked about messaging, problems solved by the message brokers, and finally, the details of the software architectures that is related or created with the message brokers. Now, we are ready to dive into the concepts of the messaging. The following screenshot describes the overall picture of messaging concepts. We have **Producers**, who are responsible for creation of messages; **Message Brokers**, who are responsible for ensuring the message sending from **Producer** to **Consumer**; **Consumers**, who are responsible for receiving the messages; and messages, who are the entity that will be sent and received.

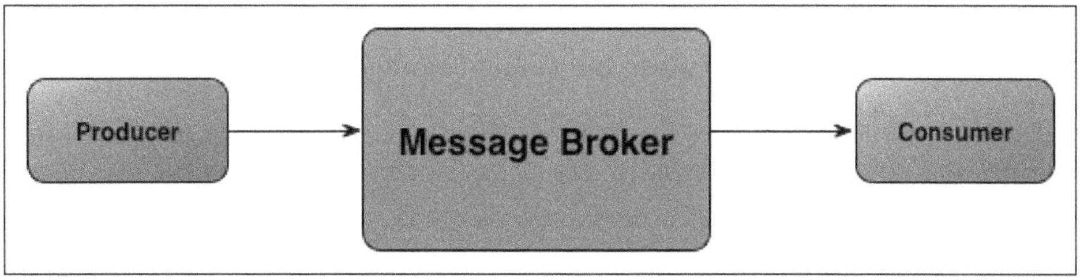

Image 8: Producer, Message Broker and Consumer

Message producers

A **message producer**, as implied the functionality by its name, produces the messages. It just sends the message to the consumer via Message Broker. Mainly, software applications, modules, and processes form producers.

A producer doesn't have a limit for messages, so it can send a large number of messages. Moreover, it clarifies the receiver. However, routing to the right receiver is not the producer's duty. Finally, one of the duties of the producer is to specify the Message Broker's network address such as IP address and its port address.

Consequently, Producer has the least duty in the messaging concepts. It has mainly one duty that is sending the message.

Message brokers

Message Brokers are the managers of messaging. They have heavy duties, such as routing the message to the right queue, controlling the size of queues, and ensuring the message sending. Thus, Message Brokers control and manage the messaging activity between producers and consumers.

Message Brokers behave like middleware between producers and consumers. Therefore, many software engineers called the message broker's place in the software system architecture as Message Oriented Middleware. As discussed before, Message Oriented Middleware solves some software engineering problems such as interoperability, enterprise integration of software systems, abstraction, and so on.

The functionalities of the Message Brokers are many, but it is good to list the most important ones:

- Ensuring the message sending to the right receiver
- Routing the message to the right queue and right receiver
- Supporting the different routing algorithms such as **Pub-Sub**, **Direct**, **Topic based**, and so on

- Scaling and the Queues
- Providing the temporary and permanent storage to the messages

As a result, Message Brokers are the brain of the messaging system. So, they are the most important concept of the system. Message Brokers affect all of the subsystems of the messaging system

Message consumers

Producers send message and Message Brokers manage the messaging functionality and routes message to the right consumer. Finally, consumer has one main responsibility which is receiving messages or listening to messages in messaging terms. Consumers await the upcoming messages, then process the message into the meaningful format and use it.

Consumers are the last point of the messaging systems. Therefore, every message has recipient or recipients, which are formed by the consumers. As we exactly determined the functions of the consumers, we should now talk about how consumers perform this listening functionality. As we know that message brokers are able to satisfy both synchronous system and asynchronous system, consumers can listen to the messages in a blocking or non-blocking way. Blocking the complete system is not preferable for all software systems; it is always good to implement asynchronous system (non-blocking) if possible. We will talk about both synchronous and asynchronous ways in the client chapters starting from *Chapter 9, Java RabbitMQ Client Programming*.

Consequently, consumers are the end point of the messaging system. They have the ability to listen to messages in both asynchronous and synchronous way. We are now ready to talk about the last concept in messaging, that is, Messages.

Messages

Messages are the main entity in the messaging systems. Producers can send messages, Message Brokers then process the messages and route the messages through queues, and lastly, consumers listen to the messages. So, every other concept in messaging operates with messages.

Messages have some information in them. Messages have headers, which have information about the sender, receiver, and message format. Moreover, messages have bodies, which have the exact information that producers send to the consumers. Message bodies could be in different types of formats such as XML, JSON, binary data, and so on.

In conclusion, Messages are fundamental to messaging system. They are container entities in the information flow.

Advanced Message Queuing Protocol (AMQP)

AMQP is the abbreviation of Advanced Message Queuing Protocol. AMQP creates the interoperability between Producer, Message Broker, and Consumer. First of all, we need to answer this question: Why we need AMQP? Since different types of message formats and different types of routing formats need to be standardized, AMQP organization creates a well-defined industry-wide messaging middleware standard.

As we discussed, AMQP's main responsibility is the interoperability of the systems inside the messaging systems. Therefore, we need to explain the scope of the AMQP as explained in the *AMQP Specification Document*:

- A defined set of messaging capabilities
- A network wire-level protocol

AMQP organization created the standard with the help of the requirements from well-known companies such as Cisco Systems, JPMorgan Chase, Red Hat, and so on. The most important requirements of the Advanced Message Queuing Model are listed as follows, taken from the AMQP Specification Document:

- To guarantee interoperability
- To provide explicit control
- To allow complete configuration

Moreover, AMQP organization clarifies the requirements of the AMQP transport layer as follows, taken from the AMQP Specification Document:

- To use binary encoding
- To handle messages
- To be long-lived
- To allow asynchronous system
- To be easily extended

From now on, we will dive into the details of the AMQP, starting with AMQ Model Architecture and its elements.

AMQ elements

As we talked before, AMQ stands for Advanced Message Queuing, and we are now talking about the elements of AMQ and its main architecture. We can express the main architecture of the middleware as follows: producer/publisher creates or sends messages; then, messages arrive at Exchanges; after that, messages are routed through the Message Queues with related Bindings to the right consumer. So, we have four model elements:

- **Message Flow**: It explains the message life cycle
- **Exchanges**: It accepts messages from publisher, and then routes to the Message Queues
- **Message Queues**: It stores messages in memory or disk and delivers messages to the consumers
- **Bindings**: It specifies the relationship between an exchange and a message queue that tells how to route messages to the right Message Queues

We are now ready to talk about the details of each AMQ Element. In addition, you can find the well-defined AMQ elements in the following screenshot:

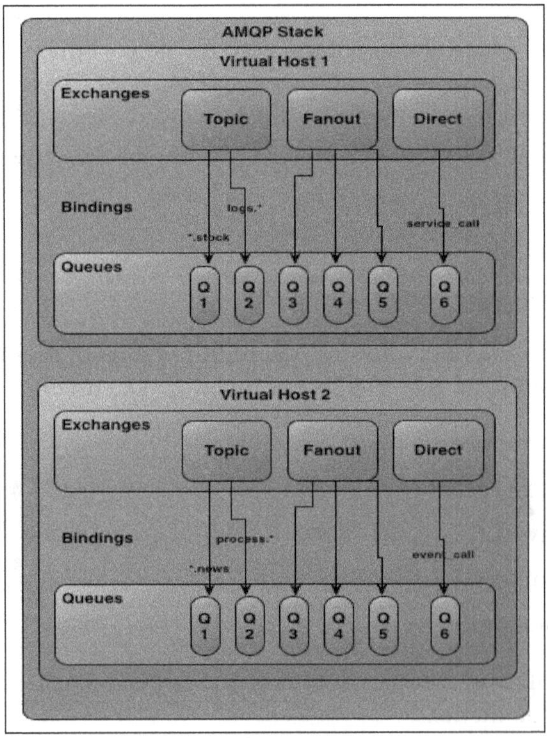

AMQP Stack

Message flow

In a nutshell, Message Flow starts when the Producer creates message and sends it to the Exchange. Then, Exchange routes to the related Message Queue with given Bindings. Finally, Consumer receives the sent message. The well-defined explanation of the Message Flow is listed as follows:

1. **Message**: This is produced by the Publisher application using AMQP Client with placing related information such as Content, Properties, and Routing Information to the Message.
2. **Exchange**: This receives the Message, which is sent from the Producer, then routes message to the right Queues, which is set on the message's Routing Information. Message will be sent to multiple queues, since it is determined with the Bindings.
3. **Message Queue**: This receives the Message and adds it to their waiting list. As soon as possible, Message Queue sends message to the related consumer. If Message Queue cannot send the Message, it stores the Message in a disk or memory.
4. **Consumer**: This receives the Message and sends Acknowledgement Message (usually it is sent automatically) to the Publisher.

You can find the well-defined Message Flow in the following screenshot:

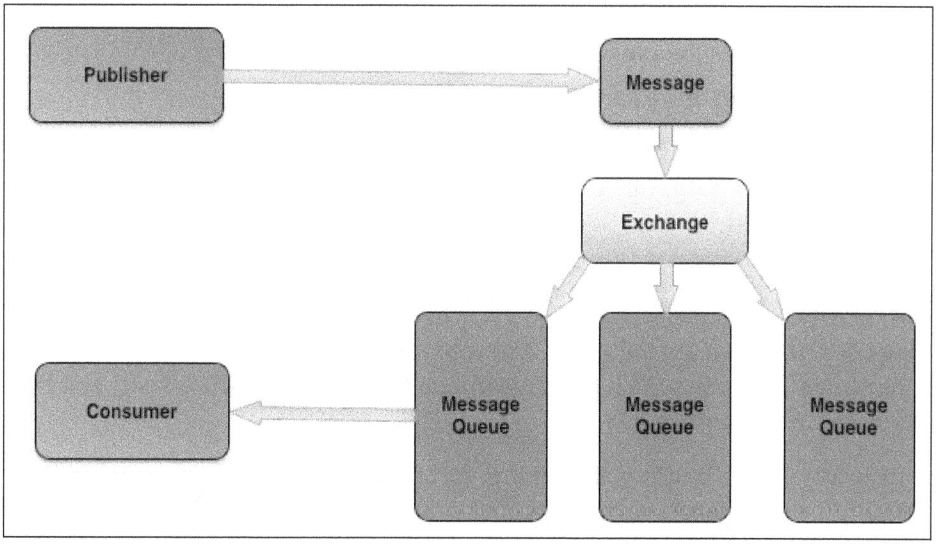

Message Flow

Exchanges in AMQ

Exchanges generally take message and route it into zero or more message queues. The routing algorithm can be determined with the bindings, which is well discussed in the *Functional Specification* topic in this chapter.

Exchanges are declared with following important properties:

- **Name**: Usually, server gives its name automatically
- **Durable**: Message Queue remains present or not, depending on whether durable is set or transient is set
- **Auto-delete**: When all queues finish, exchanges are deleted automatically

Message queues

Message Queue in AMQ is similar to the other messaging systems or task queuing systems. They store the messages in a **First-In-First-Out** (**FIFO**) way that is well defined in the queue data structure. Different from Queue data structure, if multiple readers from a queue is active, then one of the reader sometimes has a priority over another. Then, prior one takes the message before the other readers. Therefore, message queue in AMQ model is called as **weak-FIFO**.

Message Queues have the properties like Exchanges. The most important ones are listed as follows:

- **Name**: Defines the name of Message Queue
- **Durable**: If set, the Message Queue can't lose any message
- **Exclusive**: If set, the Message Queue will be deleted after connection is closed
- **Auto-delete**: If set, the Message Queue is deleted after last consumer has unsubscribed

Bindings

Bindings are rules that Exchanges use to route messages between message queues. Thus, bindings clarify in which message queue the message will be sent. The binding is determined with **routing key**.

As an example of the Bindings in real life, for instance, you have three different ways to go to your favorite restaurant, and you have to decide one of the ways. The decision is determined with the help of Bindings.

AMQ supports different type of Bindings, which will be discussed in the next section.

Functional specifications of AMQP

After defining each AMQ elements, we are now ready to express functions of each element. As a brief introduction, Messages are the main element of the system, Virtual Hosts is a way to execute more than one RabbitMQ Server instance in a server, and Exchanges routes the messages. Let's dive into the functional details of each element.

AMQP messages

Message is the main entity of messaging system as well as AMQP. It is the atomic unit of processing of the middleware routing and queuing system according to AMQP specification.

A Message consists of these following attributes:

- Content that is a binary data
- Header
- Properties

The following screenshot gives the general idea of the AMQP Message:

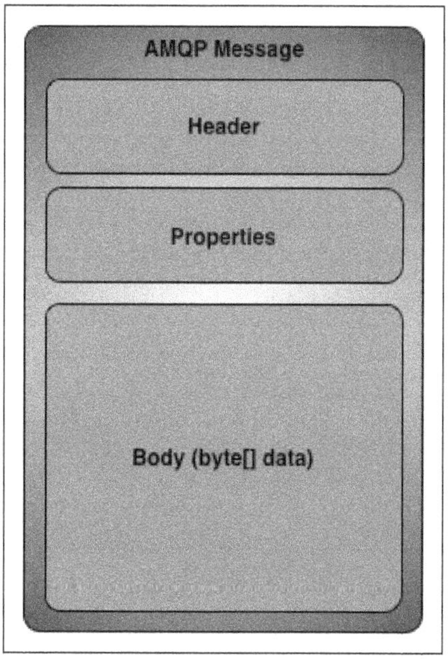

AMQP Message

Because of storing binary data on AMQP's content, AMQP has the capability to transfer file, creating application level message, and data streaming such as video streaming. Messages may be persistent if set, and may have priority level, which gives high priority messages to be sent ahead of lower priority messages waiting in the same message queue.

Virtual hosts

AMQP has a functionality to have multiple isolated environments, which have groups of users, exchanges, message queues, and so on with the help of Virtual Hosts. It is really similar to the Virtual Hosts of any Web Server in the enterprise.

Clients have an option to select a Virtual Host from the Virtual Host list. The command line tool `rabbitmqctl` manages Virtual Hosts. Authorization mechanism of each Virtual Host could be different. Clients have to choose one of the Virtual Hosts, since a Client cannot be allowed to connect to another Virtual Host while connected to one Virtual Host.

Exchange types

As explained in the AMQ Elements, Exchanges is a message routing agent within a Virtual Host. Exchanges receive the messages and route to the zero or more Message Queues. Exchanges have properties, which is well defined in the previous topic. The routing algorithm is determined using exchange type. We have five exchanges types in AMQP. Note that these are the types by default, but you can extend AMQP and create your own type of exchange. Exchange Types with their functionalities are listed as follows.

The direct exchange type – amq.direct

The flow of direct exchange type (as shown in the following screenshot) is as follows:

1. A message queue binds to the exchange using a routing key, K.
2. Then, a publisher sends the Exchange a message with the routing key, R.
3. The message is passed to the message queue if K equals to R.

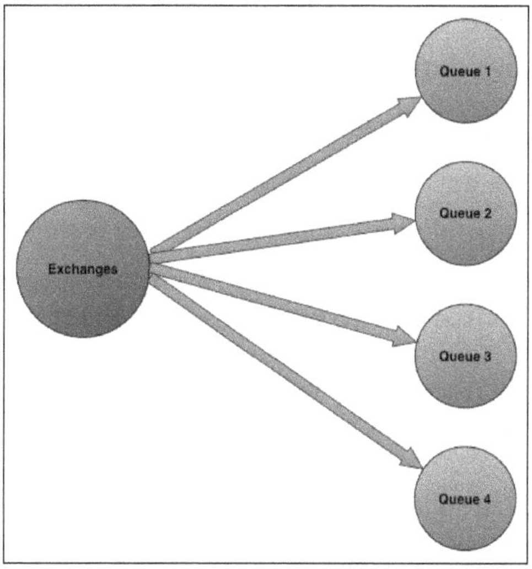

Direct Exchange Routing

The fan-out exchange type – amq.fanout

The flow of direct exchange type(as shown in the previous screenshot) is as follows:

1. A message queue binds to the exchange with no arguments.
2. Whenever a publisher sends the Exchange a message, the message is passed to the message queues unconditionally:

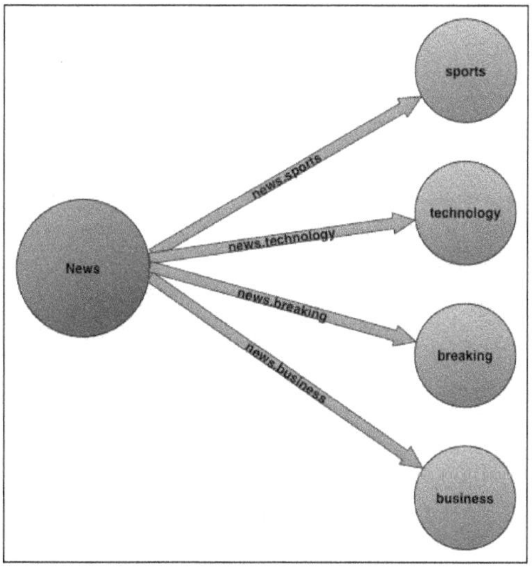

Fanout Exchange Routing

The topic exchange type – amq.topic

The flow of direct exchange type (as shown in the previous screenshot) is as follows:

1. A message queue binds to the Exchange using a routing pattern, P.
2. A publisher sends the exchange a message with the routing key, R.
3. The message is passed to the message queue if R matches P.
4. Matching algorithm works as follows: The routing key used for a topic exchange must consist of zero or more words delimited by dots such as "news.tech". The routing pattern works like a regular expression such as "*" matches single word and # matches zero or more words. For instance, "news.*" matches the "news.tech".

The headers exchange type – amq.match

Headers Exchange Type is the most powerful exchange type in AMQP. Headers exchange route messages based on the matching message headers. Exchange ignores the routing key. Whenever creating the exchanges, we specify the related headers on the exchanges, so message's headers are matched with the exchange headers using "x-match" argument. We will be looking at this Exchange Type in the Client Chapters.

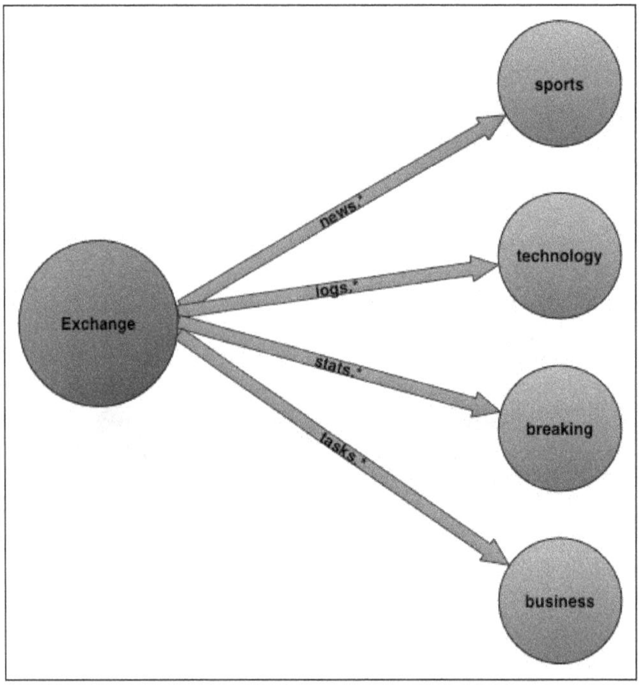

Topic Exchange Routing

Summary

In conclusion, Messaging and Message Brokers are able to solve today's problems of software engineering such as Interoperability, Heterogeneous Integration, Scalability, and so on. Moreover, Message Brokers give us high level software architectures to make our software systems more abstract.

Today, we have more complicated software systems. Moreover, we have to combine different software systems. Messaging systems is just what we need in our integration of different software systems.

We learned that Messaging has four main components: Producer, Message Broker, Consumer, and the Message. Message Broker was our manager of the messaging system; however, we had to define its standards and functionalities. Then it came to AMQP. AMQP just clarifies how Message Broker works, how clients (consumers and producers) talk with the server, and how each components of the AMQP interact with each other.

From now on, we are starting the functionalities of the RabbitMQ Server, beginning with *Chapter 4, Clustering and High Availability*.

4
Clustering and High Availability

Dan Kegel published his well-known problem, **C10K** in 1999. The problem simply arose from handling 10k simultaneous clients on the web servers. Currently, we have to handle more than 100k simultaneous clients on our web servers or on our software systems.

C10K is a great start to solve the scalability problem; however, we have a much bigger problem on our hands now. If we return to messaging systems and RabbitMQ, we have to handle lots of simultaneous messages; however, we don't have a chance to handle all simultaneous messages in a single RabbitMQ server.

Anyway, RabbitMQ has great skills to handle lots of messages in a single machine, such as more than 50k messages per second according to *VMware Performance Bookmarks*; however, as we said earlier, we need more than that. So we have to use multiple RabbitMQ servers. As a result, we need to create clusters of the RabbitMQ server to handle lots of messages per second. High availability is directly related to the scalability issue. As we make progress on the performance of the RabbitMQ using the clusters, we enhance the availability of the messaging system. Furthermore, RabbitMQ gives us a chance to control the queues for high availability.

Let's briefly define concepts that we will discuss here. High availability is generally regarded as any component or system that is operational for a specified length of time. High reliability means that a system performs consistently. A cluster is a group of computers that tries to solve the same problem unanimously.

We will cover the following topics and their solutions in this chapter:

- High reliability in RabbitMQ
- Federation in RabbitMQ
- Clustering in RabbitMQ
- Clustering settings of RabbitMQ
- High availability of queues

High reliability in RabbitMQ

As we declared the general problems of running a single instance of RabbitMQ server on a single server, it gives us another chance to solve this problem using different ways of distribution. One of the ways is **Federation**, which simply means the transmission of messages between brokers. Another way to solve this problem is clustering, that is, running multiple nodes of RabbitMQ with coordination. Both ways have different advantages and disadvantages. Clustering is done naturally in RabbitMQ servers; however, federation needs a plugin of RabbitMQ to interact between its servers. Before diving into federation in RabbitMQ, let's talk a little bit about **Shovel**. Shovel allows configuring a number of shovels which act like a client application. A shovel connects to its source and destination, reads and writes messages, and it also handles connection failures. Shovel needs a plugin to run:

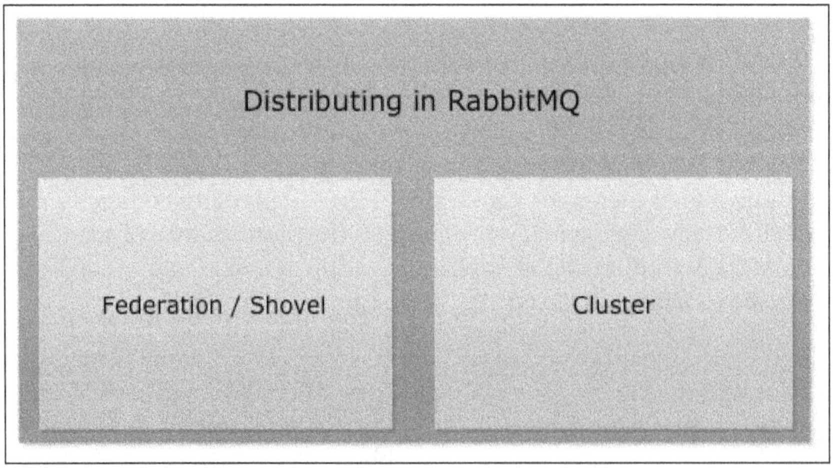

Distributing in RabbitMQ

Federation in RabbitMQ

Federation is one of the powerful ways of handling lots of messages while using multiple RabbitMQ servers. Going by the declaration from the RabbitMQ website, the main goal of Federation is to transmit messages between brokers without the need of clustering. Now, we should answer the question, why do we need federation? The following are the main reasons:

- Loose coupling
- WAN-friendly
- Scalability
- Specificity

With specificity, a broker can contain federated and local-only components. You don't need to federate everything. If you don't want to federate, you can leave it.

The `Federation` plugin is available with the standard RabbitMQ server installation. You can enable the `Federation` plugin using the following command:

```
rabbitmq-plugins enable rabbitmq_federation
```

Moreover, if you use the `management` plugin of the RabbitMQ server, you have a chance to monitor the federation using the same management plugin using the following command:

```
rabbitmq-plugins enable rabbitmq_federation_management
```

Federation-related information and configuration will be stored in the RabbitMQ database. Three levels of configuration are involved in federation according to the RabbitMQ website:

- **Upstreams**: This defines how to connect to another RabbitMQ
- **Upstream sets**: This sets the upstream groups
- **Policies**: This is a set of rules of the Federation

We can control the Federation using the management console. For instance, we can define an upstream using the following command:

```
rabbitmqctl set_parameter federation-upstream my-upstream
\'{"uri":"amqp://localhost","expires":72000}'
```

Moreover, you can set the policies of Federation using the following command of the management console:

```
rabbitmqctl set_policy --apply-to exchanges federate-me "^amq\."
\'{"federation-upstream-set":"all"}'
```

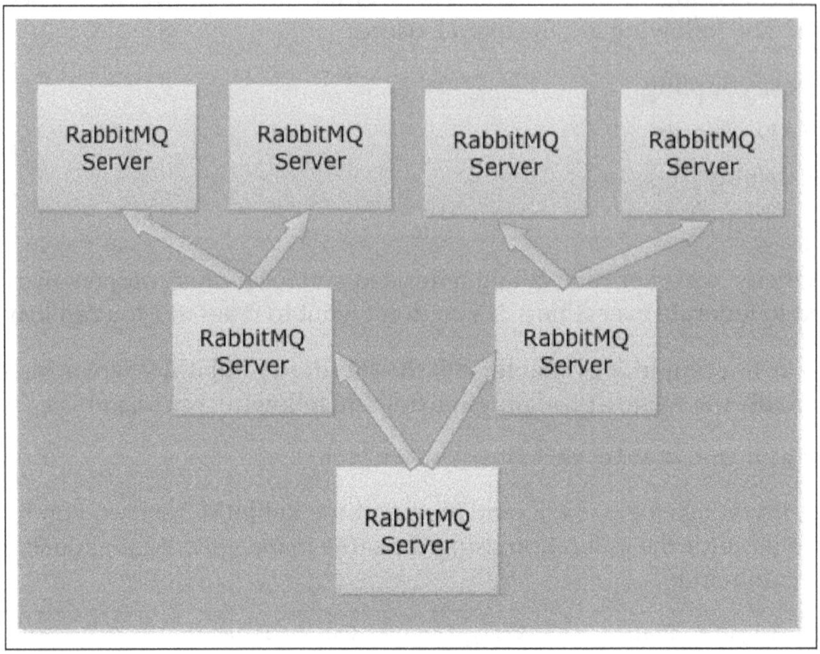

Federation in RabbitMQ

Clustering in RabbitMQ

Clustering is our main solution for handling client requests over the server applications. The RabbitMQ server also gives us cluster mechanism. Cluster mechanism replicates all the data/states across all the nodes for reliability and scalability. The general structure of the clusters would be changed dynamically, according to the addition or removal of any clusters from the systems. Furthermore, RabbitMQ tolerates the failure of each node.

Nodes should choose one of the Node type that affect the storage place; these are disk nodes or RAM nodes. If an administrator chooses a RAM node, RabbitMQ stores its state in memory. However, if an administrator chooses to store its state in a disk, then RabbitMQ stores its state on both, memory and disk.

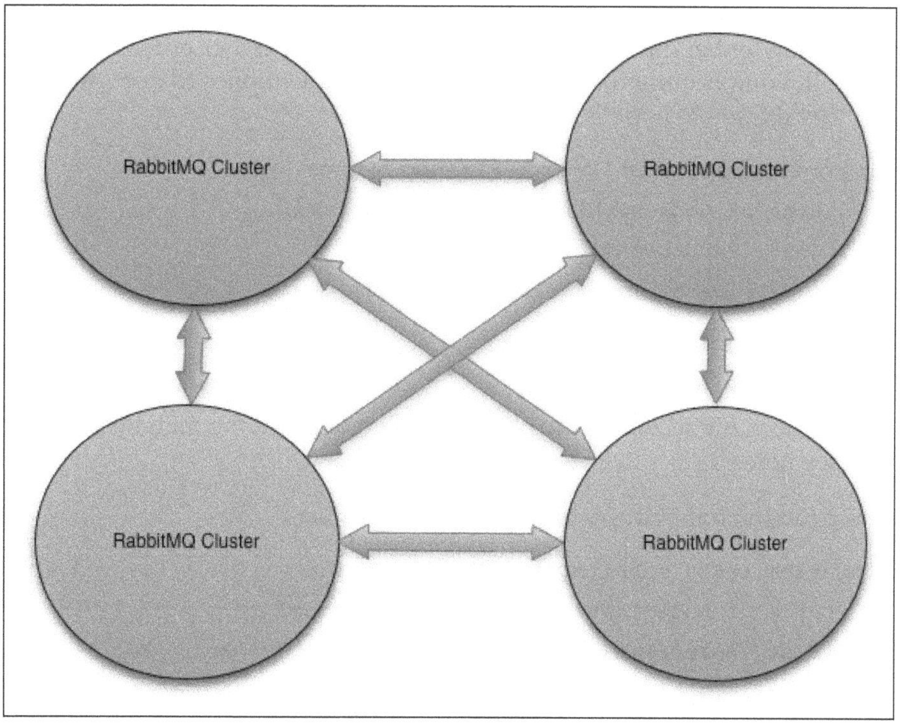

Clustering in RabbitMQ

Creating clusters

After describing the clustering in detail, we are now ready to create clusters of RabbitMQ servers on our system. Firstly, we need to start each of our RabbitMQ servers within the nodes using the following commands:

```
mastering-rabbitmq1$ rabbitmq-server -detached
mastering-rabbitmq2$ rabbitmq-server -detached
mastering-rabbitmq3$ rabbitmq-server -detached
mastering-rabbitmq4$ rabbitmq-server -detached
```

Then, with the help of `cluster_status` attribute of management console, we are ready to check whether our single node of cluster has started or not. As you can see in the following command, the management console replies to our command, provided that the single node is running:

```
mastering-rabbitmq1$ rabbitmqctl cluster_status
Cluster status of node rabbit@mastering-rabbitmq1 ...
[{nodes,[{disc,[rabbit@mastering-rabbitmq1]}]},
 {running_nodes,[rabbit@mastering-rabbitmq1]},
 {partitions,[]}]
...done.
```

We can also check the second, third, and fourth nodes with the management console as we can see here.

For Node 2 run the following command on the console:

```
mastering-rabbitmq2$ rabbitmqctl cluster_status
Cluster status of node rabbit@mastering-rabbitmq2 ...
[{nodes,[{disc,[rabbit@mastering-rabbitmq2]}]},
 {running_nodes,[rabbit@mastering-rabbitmq2]},
 {partitions,[]}]
...done.
```

For Node 3 run the following command on the console:

```
mastering-rabbitmq3$ rabbitmqctl cluster_status
Cluster status of node rabbit@mastering-rabbitmq3 ...
[{nodes,[{disc,[rabbit@mastering-rabbitmq3]}]},
 {running_nodes,[rabbit@mastering-rabbitmq3]},
 {partitions,[]}]
...done.
```

For node 4 run the following command on the console:

```
mastering-rabbitmq4$ rabbitmqctl cluster_status
Cluster status of node rabbit@mastering-rabbitmq4 ...
[{nodes,[{disc,[rabbit@mastering-rabbitmq4]}]},
 {running_nodes,[rabbit@mastering-rabbitmq4]},
 {partitions,[]}]
...done.
```

After checking the status of each node, we should go on with adding the first cluster onto another cluster. For doing this, we should first stop the application and then join clusters through the management console. Finally, we should start the application again to start the clusters:

```
mastering-rabbitmq2$ rabbitmqctl stop_app
Stopping node rabbit@mastering-rabbitmq2 ...done.


mastering-rabbitmq2$ rabbitmqctl join_cluster rabbit@mastering-

rabbitmq1


Clustering node rabbit@mastering-rabbitmq2 with [rabbit@mastering-
rabbitmq1] ...done.


mastering-rabbitmq2$ rabbitmqctl start_app
Starting node rabbit@mastering-rabbitmq2 ...done.
```

Checking the cluster status

After joining and starting the node, we can check the status of each node to see if nodes are joined to clusters or not using the `cluster_status` parameter of `rabbitmqctl` as shown by the following command:

```
mastering-rabbitmq1$ rabbitmqctl cluster_status
Cluster status of node rabbit@mastering-rabbitmq1 ...
[{nodes,[{disc,[rabbit@mastering-rabbitmq1]}],
{disc,[rabbit@mastering-rabbitmq2]}]},
},{running_nodes,[rabbit@mastering-rabbitmq1,rabbit@mastering-
rabbitmq2]},{partitions,[]}]
...done.


mastering-rabbitmq2$ rabbitmqctl cluster_status
Cluster status of node rabbit@mastering-rabbitmq2 ...
[{nodes,[{disc,[rabbit@mastering-rabbitmq2]}],
{disc,[rabbit@mastering-rabbitmq1]}]},
},{running_nodes,[rabbit@mastering-rabbitmq2,rabbit@mastering-
rabbitmq1]},{partitions,[]}]
...done.
```

A third node can also be joined to the other clusters in the same way.

```
mastering-rabbitmq3$ rabbitmqctl stop_app
Stopping node rabbit@mastering-rabbitmq3 ...done.

mastering-rabbitmq3$ rabbitmqctl join_cluster rabbit@mastering-
rabbitmq1
Clustering node rabbit@mastering-rabbitmq3 with [rabbit@mastering-
rabbitmq1] ...done.

mastering-rabbitmq3$ rabbitmqctl start_app
Starting node rabbit@mastering-rabbitmq3 ...done.
```

Now, we have three clusters. We will check each of these clusters through the management console of the RabbitMQ server:

```
mastering-rabbitmq1$ rabbitmqctl cluster_status
Cluster status of node rabbit@mastering-rabbitmq1 ...
[{nodes,[{disc,[rabbit@mastering-rabbitmq1]}],
{disc,[rabbit@mastering-rabbitmq2]},{disc,[rabbit@mastering-
rabbitmq3]}]}},{running_nodes,[rabbit@mastering-
rabbitmq1,rabbit@mastering-rabbitmq2, rabbit@mastering-
rabbitmq3]},{partitions,[]}]
...done.

mastering-rabbitmq2$ rabbitmqctl cluster_status
Cluster status of node rabbit@mastering-rabbitmq2 ...
[{nodes,[{disc,[rabbit@mastering-rabbitmq2]}],
{disc,[rabbit@mastering-rabbitmq1]},{disc,[rabbit@mastering-
rabbitmq3]}]}},{running_nodes,[rabbit@mastering-
rabbitmq2,rabbit@mastering-rabbitmq1, rabbit@mastering-
rabbitmq3]},{partitions,[]}]
...done.

mastering-rabbitmq3$ rabbitmqctl cluster_status
Cluster status of node rabbit@mastering-rabbitmq3 ...
[{nodes,[{disc,[rabbit@mastering-rabbitmq3]}],
{disc,[rabbit@mastering-rabbitmq1]},{disc,[rabbit@mastering-
rabbitmq2]}]}},{running_nodes,[rabbit@mastering-
rabbitmq3,rabbit@mastering-rabbitmq1, rabbit@mastering-
rabbitmq2]},{partitions,[]}]
...done.
```

In the same manner, the fourth and the last cluster of our software system can be joined to the clusters. After joining the last one, we are now ready to check whether all of the are joined or not.

The following code shows how to join one cluster to another cluster and monitor their status on the different clusters:

```
mastering-rabbitmq4$ rabbitmqctl stop_app
Stopping node rabbit@mastering-rabbitmq4 ...done.


mastering-rabbitmq4$ rabbitmqctl join_cluster rabbit@mastering-
rabbitmq1
Clustering node rabbit@mastering-rabbitmq4 with [rabbit@mastering-
rabbitmq1] ...done.


mastering-rabbitmq4$ rabbitmqctl start_app
Starting node rabbit@mastering-rabbitmq4 ...done.


mastering-rabbitmq1$ rabbitmqctl cluster_status
Cluster status of node rabbit@mastering-rabbitmq1 ...
[{nodes,[{disc,[rabbit@mastering-rabbitmq1]}],
{disc,[rabbit@mastering-rabbitmq2]},{disc,[rabbit@mastering-
rabbitmq3]},{disc,[rabbit@mastering-
rabbitmq4]}]}},{running_nodes,[rabbit@mastering-
rabbitmq1,rabbit@mastering-rabbitmq2, rabbit@mastering-
rabbitmq3,rabbit@mastering-rabbitmq4]},{partitions,[]}]
...done.


mastering-rabbitmq2$ rabbitmqctl cluster_status
Cluster status of node rabbit@mastering-rabbitmq2 ...
[{nodes,[{disc,[rabbit@mastering-rabbitmq2]}],
{disc,[rabbit@mastering-rabbitmq1]},{disc,[rabbit@mastering-
rabbitmq3]},{disc,[rabbit@mastering-
rabbitmq4]}]}},{running_nodes,[rabbit@mastering-
rabbitmq2,rabbit@mastering-rabbitmq1, rabbit@mastering-
rabbitmq3,rabbit@mastering-rabbitmq4]},{partitions,[]}]
...done.


mastering-rabbitmq3$ rabbitmqctl cluster_status
Cluster status of node rabbit@mastering-rabbitmq3 ...
```

```
[{nodes,[{disc,[rabbit@mastering-rabbitmq3]}],
{disc,[rabbit@mastering-rabbitmq1]},{disc,[rabbit@mastering-
rabbitmq2]},{disc,[rabbit@mastering-
rabbitmq4]}]}},{running_nodes,[rabbit@mastering-
rabbitmq3,rabbit@mastering-rabbitmq1, rabbit@mastering-
rabbitmq2,rabbit@mastering-rabbitmq4]},{partitions,[]}]
...done.


mastering-rabbitmq4$ rabbitmqctl cluster_status
Cluster status of node rabbit@mastering-rabbitmq4 ...
[{nodes,[{disc,[rabbit@mastering-rabbitmq4]}],
{disc,[rabbit@mastering-rabbitmq1]},{disc,[rabbit@mastering-
rabbitmq2]},{disc,[rabbit@mastering-
rabbitmq3]}]}},{running_nodes,[rabbit@mastering-
rabbitmq4,rabbit@mastering-rabbitmq1, rabbit@mastering-
rabbitmq2,rabbit@mastering-rabbitmq3]},{partitions,[]}]
...done.
```

Changing the cluster node types

As we specified the details of the cluster node types, we can also change the node type with the help of the management console. The `change_cluster_node_type` attribute of the management console helps us to change its node type from RAM to disk or disk to RAM. Before changing the attribute, we need to stop the application and then start the application again.

The following command shows can be used to change the type of storage of the cluster:

```
mastering-rabbitmq3$ rabbitmqctl stop_app
Stopping node rabbit@mastering-rabbitmq3 ...done.


mastering-rabbitmq3$ rabbitmqctl change_cluster_node_type ram
Turning rabbit@mastering-rabbitmq2 into a ram node
...done.
```

RabbitMQ clusters, which are joined, are able to stop on their own. This wouldn't be affected by the other clusters. Moreover, the nodes automatically catch up with the running nodes when they start up. You can check this high functionality using the management console and stopping and starting a node of the RabbitMQ server:

```
mastering-rabbitmq3$ rabbitmqctl start_app
Starting node rabbit@mastering-rabbitmq3 ...done.

mastering-rabbitmq3$ rabbitmqctl stop_app
Stopping node rabbit@mastering-rabbitmq3 ...done.

mastering-rabbitmq4$ rabbitmqctl cluster_status
Cluster status of node rabbit@mastering-rabbitmq4 ...
[{nodes,[{disc,[rabbit@mastering-rabbitmq4]}], {disc,[rabbit@mastering-
rabbitmq1]},{disc,[rabbit@mastering-
rabbitmq2]},{disc,[rabbit@mastering-
rabbitmq3]}]}},{running_nodes,[rabbit@mastering-
rabbitmq4,rabbit@mastering-rabbitmq1, rabbit@mastering-
rabbitmq2]},{partitions,[]}]
...done.
```

Updating cluster nodes

Sometimes, we need to remove the cluster from the other clusters because of the failure of the provided cluster or any other problem occurring on it. In such scenarios, we simply remove this node from the other clusters.

With the help of the management console, we are able to remove the provided node with the given `reset` attribute of console. In the following commands, you can find a related example, which resets the RabbitMQ server to remove node from the clusters.

```
mastering-rabbitmq4$ rabbitmqctl stop_app
Stopping node rabbit@mastering-rabbitmq4 ...done.

mastering-rabbitmq4$ rabbitmqctl reset
Resetting node rabbit@mastering-rabbitmq4
...done.
```

```
mastering-rabbitmq4$ rabbitmqctl start_app
Starting node rabbit@mastering-rabbitmq4 ...done.

mastering-rabbitmq1$ rabbitmqctl cluster_status
Cluster status of node rabbit@mastering-rabbitmq1 ...
[{nodes,[{disc,[rabbit@mastering-rabbitmq1]}],
{disc,[rabbit@mastering-rabbitmq2]},{disc,[rabbit@mastering-
rabbitmq3]}]}},{running_nodes,[rabbit@mastering-rabbitmq1,
rabbit@mastering-rabbitmq2]},{partitions,[]}]
...done.
```

Furthermore, we have another property of the management console that is for removing the nodes remotely. This is an amazing functionality of RabbitMQ to deal with an unresponsive node:

```
mastering-rabbitmq4$ rabbitmqctl forget_cluster_node
rabbit@mastering-rabbitmq3
Removing node rabbit@mastering-rabbitmq3 from cluster ...
...done.
```

Finally, sometimes we need to create clusters on our personal computer or we have a single instance to run the RabbitMQ server. As such, we might come across the question, how do we run multiple RabbitMQ servers on the same computer? The answer is changing the instance names and instance ports.

In the following block of code, you can find the answer to have multiple instance at one machine with the provided environment settings for port and instance names. The other operations are specified previously is as follows:

```
mastering-rabbitmq1$ RABBITMQ_NODE_PORT=5672 RABBITMQ_NODENAME=rabbit1-
node rabbitmq-server -detached
mastering-rabbitmq1$ RABBITMQ_NODE_PORT=5673 RABBITMQ_NODENAME=rabbit2-
node rabbitmq-server -detached

mastering-rabbitmq1$ rabbitmqctl -n rabbit2-node stop_app
Stopping node rabbit2-node@mastering-rabbitmq1 ...done.

mastering-rabbitmq1$ rabbitmqctl -n rabbit2-node join_cluster
rabbit@'hostname -s'
Clustering node rabbit2-node@mastering-rabbitmq1 with [rabbit1-node@
mastering-rabbitmq1] ...done.

mastering-rabbitmq1$ rabbitmqctl -n rabbit2-node start_app
Starting node rabbit2-node@mastering-rabbitmq3 ...done.
```

Clustering the settings of RabbitMQ

As we explained the joining process, resetting process, and the other processes of RabbitMQ server, all these need a change in settings. Clustering settings can be done using the management console, which changes at runtime and through the RabbitMQ `configuration` file. The following structure simply describes the settings structure of the RabbitMQ clustering.

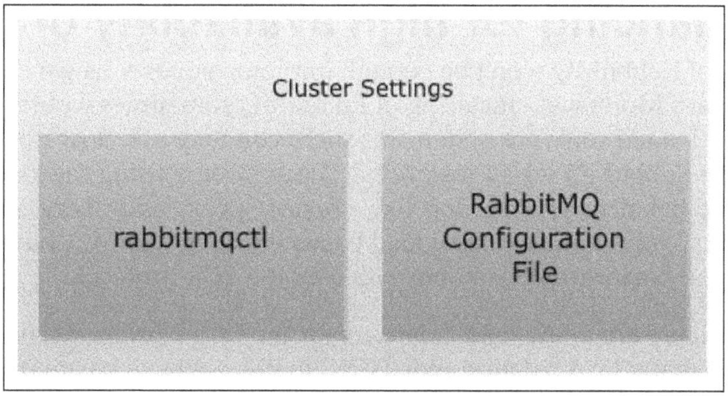

Cluster Settings in RabbitMQ

The management console attributes are well specified in the previous sections and we'd like to talk about the configuration file more. If we want to start our nodes and join them at the startup, we should add configuration parameters to the RabbitMQ `configuration` file. The `cluster_nodes` attribute of the RabbitMQ `configuration` file specifies each node of the cluster with the provided cluster node type. Furthermore, we can specify the network partition-handling mode with the provided `cluster_partition_handling` parameter of the RabbitMQ `configuration` file. Let's take a look at the following table:

Variable	Variable Description
`cluster_nodes`	If the disk type is chosen, execute the following command: `[{rabbit, [{cluster_nodes,{['rabbit@mastering-rabbitmq1', 'rabbit@mastering-rabbitmq2', 'rabbit@mastering-rabbitmq3', 'rabbit@mastering-rabbitmq4'], disc}}]}].` If the RAM type is chosen, execute the following command: `[{rabbit, [{cluster_nodes,{['rabbit@mastering-rabbitmq1', 'rabbit@mastering-rabbitmq2', 'rabbit@mastering-rabbitmq3', 'rabbit@mastering-rabbitmq4'], ram}}]}].`

Variable	Variable Description
cluster_partition_handling	This variable specifies how network partitions are handled. • Default value: ignore • Possible values: Ignore, pause_minority, or autoheal

Load balancing for high availability of queues

A single node of RabbitMQ won't be enough for your requests, as we declared in the introduction part. Moreover, clustering of RabbitMQ sometimes wouldn't be enough for your heavy loaded software system. As such, you may ask, how do we balance the loads of the RabbitMQ server instances? The answer is using the load balancing systems. A load balancing system acts like a reverse proxy and distributes the load between servers. We can make use of load balancers for RabbitMQ, but we need to use the TCP load balancers since RabbitMQ uses the TCP protocol.

In the market, there are many load balancers for the HTTP protocols; however, we haven't got too many load balancers for TCP. Finally, we have an amazing open source TCP/HTTP load balancer, that is, **HAProxy**, which is used by well-known companies. Furthermore, with the help of a module, our performance for the first web server—**Nginx**—also supports the TCP load balancing with its mechanisms.

HAProxy has its own configuration for all TCP connections as well as RabbitMQ server instances. Here, you can find the example configuration file of HAProxy for the RabbitMQ server:

(HAProxy)

```
global
log     127.0.0.1 alert
log     127.0.0.1 alert debug

defaults
log        global
mode       http
```

```
option      dontlognull
option      redispatch
retries     3
contimeout 5000
clitimeout 50000
srvtimeout 50000

listen rabbitmq 192.168.1.1:5000
mode    tcp
stats   enable
balance roundrobin
option  forwardfor
option  tcpka
server  rabbit01 192.168.1.1:5672 check inter 5000 downinter 500
server  rabbit02 192.168.1.2:5672 check inter 5000 backup
server  rabbit03 192.168.1.3:5672 check inter 5000 backup
```

Furthermore, we can balance the loads of the RabbitMQ server instances using Nginx with the module called `nginx_tcp_proxy_module`. If you'd like to use this module, you have to compile Nginx from the source code.

Here, you can find the details of the configuration of Nginx for the RabbitMQ server:

(Nginx TCP)

```
tcp {
  upstream cluster {
    # simple round-robin
    server 192.168.1.1:5672;
    server 192.168.1.2:5672;
    check interval=3000 rise=2 fall=5 timeout=1000;
  }
  server {
    listen 5672;
    proxy_pass cluster;
  }
}
```

The following image shows the architecture of the clusters behind the proxy servers, such as Nginx:

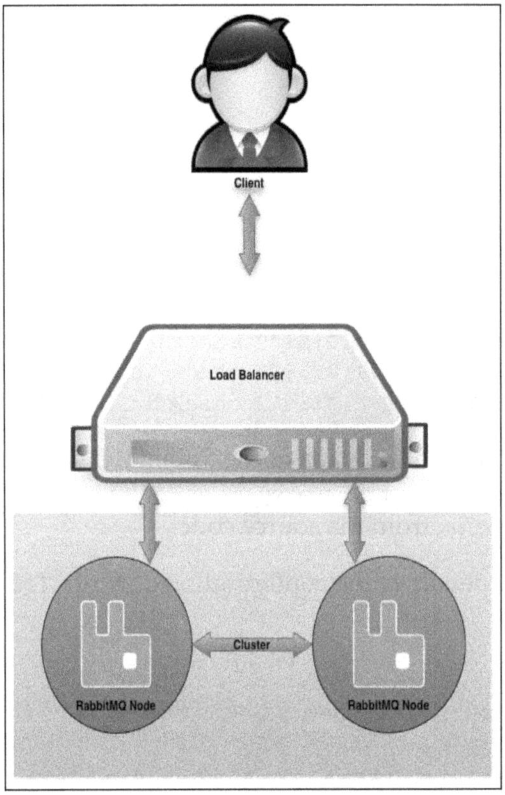

Load Balancing in RabbitMQ

Summary

Nowadays, scalability and real-time responsive software systems are our main responsibility. Therefore, all parts of our software system should fit with the other parts to sustain scalability and real-time responsivity. Moreover, the RabbitMQ server part should sustain these terms.

RabbitMQ has two types of systems for handling the lots of messages in a second: Federation and Clustering. As we explained, Federation is done by the RabbitMQ plugin; whereas, Clustering is supported with the minimal installation of RabbitMQ. Also, we should guarantee the balancing of the RabbitMQ servers. As RabbitMQ uses the TCP, we need TCP load balancers, such as HAProxy and `nginx_tcp_proxy_module` powered by Nginx. In the next chapter, we'll touch on RabbitMQ plugins and RabbitMQ plugin development.

5

Plugins and Plugin Development

Today, plugins are the main extending point of the software systems. As we know from other software projects, plugins are used to extend the capability or add any other skills to the software system.

RabbitMQ has its own plugin system, and it gives us default plugins as well. Moreover, RabbitMQ gives us one more opportunity: developing our custom plugin using RabbitMQ's API.

In this chapter, we will discuss the details of the plugins, default plugins and custom plugin development, as the following list describes:

- Plugin management and default plugins
- Plugin configuration
- Custom plugin development

Plugin management and default plugins

RabbitMQ provides a number of tools that aid us in plugin management. Additionally, RabbitMQ provides lots of default plugins to monitor, manage, add features, and so on to RabbitMQ. Plugins are crucial because they introduce an important set of new features, making RabbitMQ easier to use and manage.

Now, we are ready to talk about enabling and disabling plugins.

Enabling and disabling plugins

RabbitMQ provides a plugin management tool called **rabbitmq-plugins**. rabbitmq-plugins is a command line tool to enable, disable, and list the plugins within the RabbitMQ server. The `rabbitmq-plugins` command will enable or disable plugins by updating the plugin configuration file. It will then contact the running server to tell it to start or stop plugins as needed. You can use the `-n` option to specify a different node, or use `--offline` to only change the file.

As all management tools of RabbitMQ need write permissions, rabbitmq-plugins also needs write permissions to get allowed to be run by the user. General usage of the tool is as follows:

```
rabbitmq-plugins {command} [command param1, command param2,…]
```

The functions and the parameters of rabbitmq-plugins are listed in the following table:

Command Name	Command Parameters	Command Description
`list`	`-v`, `-m`, `-E`, `-e`, `pattern`	Lists all plugins with their versions and dependencies. `-v` parameter shows all the plugin details. `-m` parameter shows only plugin names. `-E` parameter shows only explicitly enabled plugins. `-e` parameter shows only explicitly or implicitly enabled plugins. `pattern` parameter shows only the plugins that match with the defined pattern. Example: **`rabbitmq-plugins list`** **`rabbitmq-plugins list -v`** **`rabbitmq-plugins list -m`**
`enable`	`Plugin Name1`, `Plugin Name2`	Enables the provided plugins. We can use multiple plugin names as a parameter. Example: **`rabbitmq-plugins enable rabbitmq_management`**

Command Name	Command Parameters	Command Description
`disable`	`Plugin Name1,Plugin Name2`	Disables the provided plugins. We can use multiple plugin names as a parameter. Example: **`rabbitmq-plugins disable rabbitmq_stomp`**

Table 1: `rabbitmq-plugins` commands and its parameters

The following screenshot of command line shows how to run listing command of rabbitmq-plugins:

```
jspyconf — vagrant@precise32: ~ — ssh — 115×30

vagrant@precise32:~$ rabbitmq-plugins list
[e] amqp_client                       3.0.4
[ ] cowboy                            0.5.0-rmq3.0.4-git4b93c2d
[ ] eldap                             3.0.4-gite309de4
[e] mochiweb                          2.3.1-rmq3.0.4-gitd541e9a
[ ] rabbitmq_auth_backend_ldap        3.0.4
[ ] rabbitmq_auth_mechanism_ssl       3.0.4
[ ] rabbitmq_consistent_hash_exchange 3.0.4
[ ] rabbitmq_federation               3.0.4
[ ] rabbitmq_federation_management    3.0.4
[ ] rabbitmq_jsonrpc                  3.0.4
[ ] rabbitmq_jsonrpc_channel          3.0.4
[ ] rabbitmq_jsonrpc_channel_examples 3.0.4
[E] rabbitmq_management               3.0.4
[e] rabbitmq_management_agent         3.0.4
[ ] rabbitmq_management_visualiser    3.0.4
[ ] rabbitmq_mqtt                     3.0.4
[ ] rabbitmq_old_federation           3.0.4
[ ] rabbitmq_shovel                   3.0.4
[ ] rabbitmq_shovel_management        3.0.4
[ ] rabbitmq_stomp                    3.0.4
[ ] rabbitmq_tracing                  3.0.4
[e] rabbitmq_web_dispatch             3.0.4
[ ] rabbitmq_web_stomp                3.0.4
[ ] rabbitmq_web_stomp_examples       3.0.4
[ ] rfc4627_jsonrpc                   3.0.4-git7ab174b
[ ] sockjs                            0.3.4-rmq3.0.4-git3132eb9
[e] webmachine                        1.9.1-rmq3.0.4-git52e62bc
vagrant@precise32:~$
```

Listing of Plugins

In addition, the following screenshot of command line shows how to enable or disable any plugin using the command of enable or disable of rabbitmq-plugins:

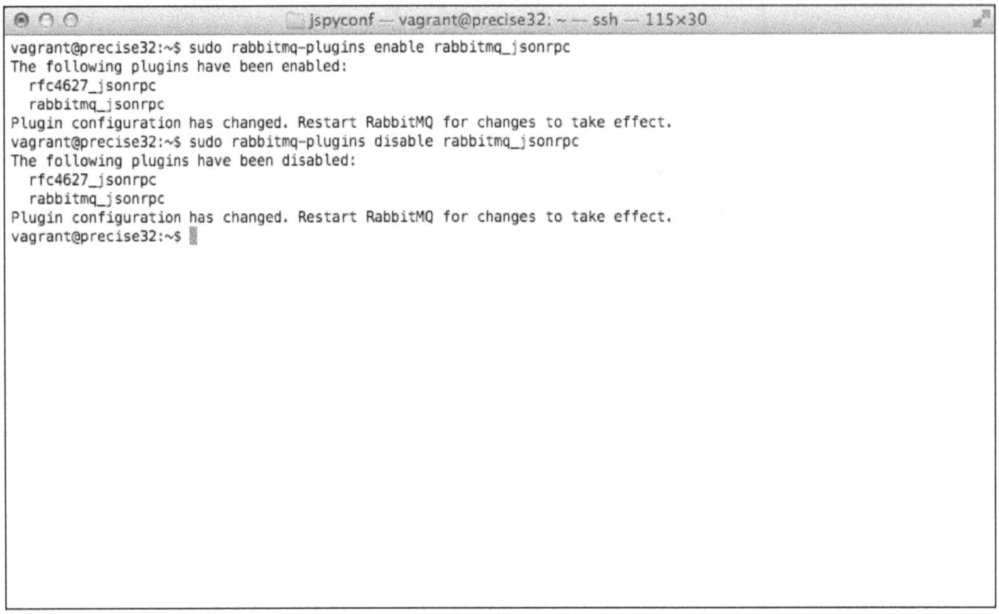
jspyconf — vagrant@precise32: ~ — ssh — 115×30
vagrant@precise32:~$ sudo rabbitmq-plugins enable rabbitmq_jsonrpc
The following plugins have been enabled:
rfc4627_jsonrpc
rabbitmq_jsonrpc
Plugin configuration has changed. Restart RabbitMQ for changes to take effect.
vagrant@precise32:~$ sudo rabbitmq-plugins disable rabbitmq_jsonrpc
The following plugins have been disabled:
rfc4627_jsonrpc
rabbitmq_jsonrpc
Plugin configuration has changed. Restart RabbitMQ for changes to take effect.
vagrant@precise32:~$

Enabling and Disabling of Plugins

Installing plugin from third-party sources

RabbitMQ gives us another opportunity to develop our own plugins using Erlang language. Moreover, lots of plugins developed by individual developers can be found on the Internet.

Now, it is time to show you how to install a plugin from source code to be enabled on the RabbitMQ Server.

Firstly, we will choose one of the open source plugins called **RabbitMQ Random Exchange Type** plugin, which can be fetched from `git://github.com/jbrisbin/random-exchange.git` via Git source code management tool using the following command:

```
git clone git://github.com/jbrisbin/random-exchange.git
cd random-exchange
make package
cp dist/*.ez $RABBITMQ_HOME/plugins
```

Then, we go to the created folder called `random-exchange`. After that, we build the source code and copy the built files to `$RABBITMQ_HOME/plugins` using the following commands:

```
make package
cp dist/*.ez $RABBITMQ_HOME/plugins
```

Now, we are ready to enable our third party plugin using the following command with the help of rabbitmq-plugins tool:

```
rabbitmq-plugins enable random-exchange
```

Default plugin list

RabbitMQ contributors publish the default plugins that will be helpful to add new protocols, some authentication functionality on RabbitMQ Server, and so on. Moreover, some experimental plugins published by the contributors are also a part of these publications.

To talk in detail regarding the plugins, let's move on to the following table that shows the default plugins and their description:

Plugin Name	Plugin Description
`rabbitmq_auth_backend_ldap`	To use external LDAP as authentication or authorization functionality. This plugin will be explained in detail in *Chapter 8, Security in RabbitMQ*.
`rabbitmq_auth_mechanism_ssl`	To add authentication mechanism using SSL certificates. This plugin will be explained in detail in *Chapter 8, Security in RabbitMQ*.
`rabbitmq_consistent_bash_exchange`	To provide the consistent hashing on the exchanges to ensure that all queues bound to exchange will receive an equal number of messages.
`rabbitmq_federation`	To transmit messages between brokers without clustering within wide area networks. Provides scalability skills. This plugin was explained in detail in *Chapter 3, Architecture and Messaging*.

Plugin Name	Plugin Description
`rabbitmq_federation_management`	To manage the federation plugin in RabbitMQ Server with provided API and UI. Federation Management plugin is active when the RabbitMQ Management plugin is also active.
`rabbitmq_management`	To monitor and manage RabbitMQ using web based applications. Applications use RabbitMQ API, one of the most useful plugins in RabbitMQ.
`rabbitmq_management_agent`	To monitor and manage the clusters of the RabbitMQ needs.
`rabbitmq_mqtt`	Enables the supporting of MQTT 3.1 protocol in RabbitMQ. MQTT is simply defined as lightweight publish/subscribe messaging transport.
`rabbitmq_shovel`	Enables the ability to continually consume messages from one queue and publish them to exchanges in another broker. This plugin's functionality was explained in *Chapter 3, Architecture and Messaging*.
`rabbitmq_shovel_management`	To monitor and manage the shovel in RabbitMQ Server with provided UI and API. Shovel Management plugin is active when the RabbitMQ Management plugin is active.
`rabbitmq_stomp`	Enables the gateway to expose the AMQP functionality using the STOMP protocol that provides an interoperable wire format.

As discussed earlier, contributors publish some experimental plugins; for instance, supporting the AMQP version 1.0 plugin. The following table shows the details of the experimental plugins and their description:

Plugin Name	Plugin Description
`rabbitmq_amqp1_0`	Enables the AMQP version 1.0 RabbitMQ Server.
`rabbitmq_jsonrpc_channel`	Enables the AMQP over HTTP protocol using `jsonrpc`. Moreover, this plugin provides JavaScript libraries to communicate with RabbitMQ.
`rabbitmq_jsonrpc_channel_examples`	To provide the examples of AMQP over HTTP, such as shared whiteboard, chat application, and some tests.
`rabbitmq_management_visualizer`	Visualizes the broker topology using the management web application by adding the **Visualizer** tab.
`rabbitmq_tracing`	Enables message tracing and logging.
`rabbitmq_web_stomp`	To attach rabbitmq_stomp to web browsers using the HTML 5 **WebSockets** layer **SockJS**.
`rabbitmq_web_stomp_examples`	To provide the `rabbitmq_web_stomp` examples, such as simple collaboration tool.

Plugin configuration

As we know from *Chapter 2, Configuring RabbitMQ*, the configuration file locates the plugin-related configuration. The following table shows the plugin-related configuration parameters in the RabbitMQ configuration file:

Name	Default Value	Description
`rabbitmq_plugins_dir`	`$RABBITMQ_HOME/plugins`	The location where plugins of RabbitMQ Server are located.
`rabbitmq_plugins_expand_dir`	`$RABBITMQ_MNESIA_BASE/$RABBITMQ_NODENAME-plugins-expand`	The location where RabbitMQ expands—plugins are zip files with `.ez` extension—enabled plugins when starting the server.

Custom plugin development

As we discussed earlier about the custom plugin development, RabbitMQ gives us a chance to develop our own plugins. Sometimes we need to access internal functionality of RabbitMQ, which is not possible with AMQP interface. Therefore, we need to design and develop our custom plugins.

Ok, we decided to develop our custom RabbitMQ plugin. Now, we should know the requirements for custom plugin development in RabbitMQ. As RabbitMQ is developed in Erlang, we have to know Erlang system and its design principles first. After that, we have to know the internal API of RabbitMQ to use APIs in plugin. To access the RabbitMQ APIs, we need a working RabbitMQ development environment. Therefore, we need to download all source code using Mercurial source code management tool and make the source code available using the following command:

```
http://hg.rabbitmq.com/rabbitmq-public-umbrella/hg clone
http://hg.rabbitmq.com/rabbitmq-public-umbrella/
cd rabbitmq-public-umbrella
make co
make
```

Output of the public umbrella plugin is as follows:

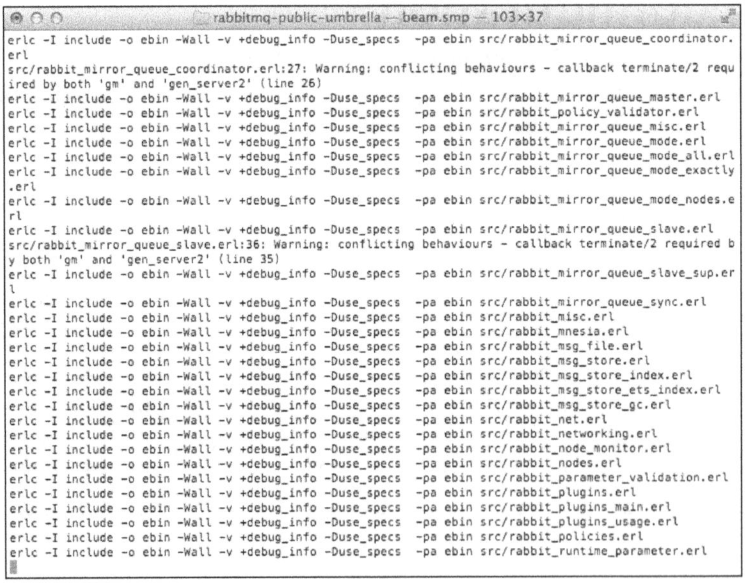

Making the RabbitMQ from Source Code

After preparing the development environment, we are now ready to talk about the basics of Erlang.

Basics of Erlang

Erlang was developed by Ericsson to manage telecom projects to support distributed, real-time, high availability applications. Erlang is a purpose oriented programming language that is concurrent and distributed naturally. The first version of Erlang was released in 1986, and the first open source version of Erlang was released in 1998. Erlang is a general-purpose, concurrent, garbage-collected programming language and runtime system.

Erlang relies on a very simple concurrency model that allows individual blocks of code to be executed multiple times on the same host. Additionally, Erlang provides a failure model on its concurrency model to handle errors on the processes. Thus, developing distributed, scalable, and highly fault tolerant software systems could be easily done by Erlang.

Erlang is a functional programming language similar to **Clojure**, **Scala**, and so on. After talking about the brief introduction of Erlang, let's now move on to the basics of Erlang.

If you installed the RabbitMQ Server on your computer with the help of *Chapter 1, Getting Started*, you also have Erlang runtime environment on your computer. If not installed, please turn back to *Chapter 1, Getting Started*, and read the installation instructions. When you run the following command on your Terminal or Command:

```
$ erl
```

You will get the following Erlang shell:

```
Erlang R16B03 (erts-5.10.4) [source] [64-bit] [smp:4:4] [async-
threads:10] [hipe] [kernel-poll:false]


Eshell V5.10.4  (abort with ^G)
1>
```

Then, if you add two integers with a stop point and press *Enter* key, you'll get the sum of these integers as follows:

```
1> 10 + 27.
37
```

Single line comments on the Erlang can be shown with the percentage (%) symbol such as:

```
% This is comment
```

Variables and expressions

Variables in Erlang are similar to any other dynamic interpreted programming languages such as Python, Ruby, and so on. Number types are separated into two, integers and float numbers, as shown in the following code:

```
2> 3+4.
7
3> 3.5 * 3.6.
12.6
```

Strings are described with the double or single quotation marks. Moreover, Erlang has lots of helper functions for strings:

```
5> "ahmet".
"ahmet"
8> string:substr("ahmet",1,3).
"ahm"
9> string:join(["one","two","three"],", ").
"one, two, three"
```

The `substr()` function takes the substring of the string (`"ahmet"`) within the given range indexes. The `Join()` function takes the string list and joins them with a separator. Erlang has lots of utility functions for String.

Erlang supports all Boolean expressions such as equality, more than, less than, and so on. Examples of the Boolean expressions are listed in the following command lines:

```
10> "abc" < "def".
true
11> "abc" == "def".
false
12> 5 == 5.
true
```

Lastly, Erlang has different variable type from other programming languages called **Atoms**. Atoms are similar with the #define value in C. Atoms are named constants and they are used for comparison. We'll see the usage of atom type variable in the *Function and Modules* section.

Tuples and lists

Tuples is the compound data type that stores the fixed number of elements. It is similar to Python's tuples. The following examples show Tuple and its utility functions that is provided by Erlang:

```
1> T = {test,32,{12,23,"emrah"}}.
{test,32,{12,23,"emrah"}}
2> element(1,T).
test
3> setelement(2,T,23).
{test,23,{12,23,"emrah"}}
4> tuple_size(T).
3
```

The `element()` function gets the related element with the given index. The `setelement()` function sets the element with new value with given index. Lastly, tuple size is showed with the `tuple_size()` function.

`List` is a compound data type that stores the variable number of elements. It is similar to Tuples; however, some properties of List give advantages over Tuples. List has head and tail structure, which are also a List data structure. Head and Tail of the List is described as follows: `[H|T]`. The following examples show the `List` and its helper functions that is provided by Erlang:

```
1> L1 = [a,2,{c,4}].
[a,2,{c,4}]
2> [H|T] = L1.
[a,2,{c,4}]
3> H.
a
4> T.
[2,{c,4}]
5> L2 = [d|T].
[d,2,{c,4}]
6> length(L1).
3
7> lists:append([L1,5]).
[a,2,{c,4}|5]
```

We first assign a list to a variable. In Erlang, after assignment of a variable, we cannot assign to the variable again. In the second example, we were splitting the list into head and tails lists. Moreover, we have the `length()` function to calculate the length of the lists. Additionally, Erlang provides lots of list utility functions, and the `append()` function is one of them. It appends an element onto the list.

Functions and modules

As we want to make our components reusable, we have to use such structures to store functions, attributes, and so on. In Erlang, we are using Modules to reuse our functional elements. A module in Erlang consists of attributes and function declarations. The following example shows the Fibonacci series function within module code:

```
%File Name: fact.erl
-module(fact).          % module attribute
-export([fact/1]).      % export attribute
fact(N) when N>0 ->     % beginning of function declaration
N * fact(N-1);
fact(0) ->
1.                      % end of function declaration
```

As we look at *fibo.erl* in detail, we can see the module attributes, module, export, and functions with their statements. The `module` attribute gives the name to the module that will be useful when you import the module. The `export` attribute defines each function with their length of parameters. If we want to import this code into the Erlang shell, we can use the following:

```
1> c(fact).
{ok,fact}
2> fact:fact(5).
120
```

As you can see, we first compile the source code using the `c` function. Then, we are able to call the function fact using the `module` name (`fib.erl`). Moreover, we can call the same name function with the one parameter atom. However, functionality of the function is different. The following example code shows these functions:

```
%Filename: fib.erl
-module(fib).
-export([fib/1]).

fib(0) -> 0;
fib(1) -> 1;
fib(N) when N > 1 -> fib(N-1) + fib(N-2).
```

Finally, if we compile and run the **Fibonacci** sequence code, we'll get the following result in our command line:

```
2> c(fib).
{ok,fib}
3> fib:fib(5).
5
4> fib:fib(10).
55
```

Functions in the Erlang simply match the parameters with the provided parameters from the function users. If parameters are matched with the function, Erlang runtime calls matched function. As you can see, functions of Erlang simply fit well with the recursive algorithms. The last example code shows you how well-known Merge Sort could be written in four lines of code:

```
%Filename: sorting.erl
-module(sorting).
-export([mergeSort/1]).

mergeSort(L) when length(L) == 1 -> L;
mergeSort(L) when length(L) > 1 ->
{L1, L2} = lists:split(length(L) div 2, L),
lists:merge(mergeSort(L1), mergeSort(L2)).
```

Will give the following output:

```
2> c(sorting).
{ok,sorting}
3> sorting:mergeSort([1,34,21,22,42,55]).
[1,21,22,34,42,55]
```

Conditionals

Erlang has the if clauses; however, its structure is somehow different from other programming languages. The structure of the `if` clause can be seen in the following example:

```
%Filename comp.erl
-module(comp).
-export([compare/2]).

compare(A,B) ->
  if
```

```
        A > B ->
          a_more_than_b;
        B > A ->
          b_more_than_a;
        A == B ->
          a_is_equal_to_b
      end.
```

After running the preceding code in Erlang shell, you will get the following command line:

```
3> c(comp).
{ok,comp}
4> comp:compare(10,5).
a_more_than_b
5> comp:compare(5,5).
a_is_equal_to_b
6> comp:compare(5,7).
b_more_than_a
```

Looping in Erlang

As Erlang gives us a chance to develop our codes easily in a **recursive** way, we don't need to use `for` loops similar to other programming languages. Therefore, if we need to iterate over `list` or any other data structure, all we need to do is develop recursive function.

The following example shows how to sum up all the elements within the list:

```
%Filename: sum.erl
-module(sum).
-export([sum/1]).

sum([]) ->
  0;
sum([H|T]) ->
  H + sum(T).
```

Note that, `H` and `T` are arguments to `sum()` function.

Now, we are ready to compile and run the module in the Erlang shell:

```
12> c(sum).
{ok,sum}
13> sum:sum([1,3,2,4435,232,1]).
4674
14> sum:sum([]).
0
```

Erlang has some helper functions in its data structures. The `foreach()` function is one of the helper functions of the list data structure. The following code is one of the examples of `foreach` attribute:

```
%Filename: iter.erl
-module(iter).
-export([iter/1]).

iter(L) ->
  lists:foreach(fun
    (N) ->
      io:format("Value: ~p ",[N])
  end, L).
```

Now, we can compile and run the `foreach` code, as shown in the following example:

```
16> c(iter).
{ok,iter}
17> iter:iter([1,2,34,3,25,24]).
Value: 1 Value: 2 Value: 34 Value: 3 Value: 25 Value: 24 ok
18> iter:iter([1,6,32,32,2,34,13,67,25,24]).
Value: 1 Value: 6 Value: 32 Value: 32 Value: 2 Value: 34 Value: 13
Value: 67 Value: 25 Value: 24 ok
```

Concurrent programming

As earlier explained, one of the main reasons for developing in Erlang is its capacity to handle concurrency and distributed programming. With concurrency, we can run programs that will run in the numerous threads. Erlang gives us an amazing chance to create parallel threads and communicate these threads with each other easily. Erlang has no mutable data structures, which means no locks are need for threading. This removes a lot of the complexity while programming concurrent programs. It also means that every time you think you are changing a variable, you are actually getting a new copy of the variable with new value, and not actually changing the value.

Erlang calls the threads of execution as process. Erlang creates the new threads using the `spawn()` function. Definition of spawn function is as follows:

```
spawn (Module, Function, Arguments)
```

In the following example you can see the easy usage of creating the threads using spawn:

```
% Filename: talk.erl
-module(talk).
-export([talk/2,run_concurrently/0]).

talk(Word, 0) ->
  done;
talk(Word, N) ->
  io:format("~p~n",[Word]),
  talk(Word, N - 1).

run_concurrently() ->
  spawn(talk, talk, [hello, 5]),
  spawn(talk, talk, [world, 4]).
```

The following command line is viewed after compiling and running the module and function:

```
8> c(talk).
talk.erl:4: Warning: variable 'Word' is unused
{ok,talk}
9> talk:run_concurrently().
hello
world
hello
world
<0.74.0>
hello
world
hello
world
hello
```

Erlang also gives us the opportunity to send and receive messages between threads, using the `receive` construct. The `receive` construct has one role: to allow processes to await messages from the other threads. The structure of receive is as follows:

```
receive
  pattern1 ->
    actions1;
  pattern2 ->
    actions2;
  pattern3 ->
    actions3
end.
```

Sending message is transmitted by the operator "`!`". The syntax of "`!`" is as follows:

```
Pid ! Message
```

The following code is an example to send and receive messages between threads using the Erlang's helper message receiver structure:

```
% Filename: msg.erl
-module(msg).
-export([sender_func/2,receiver_func/0,start_func/0]).

sender_func(0, Sender_PID) ->
  Sender_PID ! finished,
  io:format("Sender is Finished~n",[]);
sender_func(N, Sender_PID) ->
  Sender_PID ! {sender_func, self()},
  receive
    receiver_func ->
      io:format("Sender received message~n",[])
  end,
  sender_func(N-1, Sender_PID).

receiver_func() ->
  receive
    finished ->
      io:format("Receiver finished~n",[]);
    {sender_func, Sender_PID} ->
      io:format("Receiver receives message~n",[]),
      Sender_PID ! receiver_func,
      receiver_func()
  end.
```

```
start_func() ->
  Receiver_PID = spawn(msg, receiver_func, []),
  spawn(msg, sender_func, [5, Receiver_PID]).
```

As you see in the preceding example code, we have two functions: `sender_func` sends the messages in a recurrent way, while `receiver_func` receives the messages and outputs them. Whenever we want to send message to the other thread, we have to send the message with the destination's PID, where PID is process identifier. Therefore, you can see the PID related information in the sending message structure. Moreover, you see that the receive structure helps receive the message inside the thread functions while filtering the message. The following command line shows the compiled message code and its functions:

```
5> c(msg).
{ok,msg}
6> msg:start_func().
Receiver receives message
<0.55.0>
Sender received message
Receiver receives message
Sender received message
Receiver receives message
Sender received message
Receiver receives message
Sender received message
Receiver receives message
Sender received message
Sender is Finished
Receiver finished
```

Simple RabbitMQ metronome plugin

Now we have nearly learned to write code in Erlang. Our final task is to develop our own RabbitMQ plugin called **Metronome**, which is an official custom plugin of the RabbitMQ. It is published at `https://www.rabbitmq.com/plugin-development.html`.

Metronome plugin simply declares an exchange called "metronome" and sends a message every second with routing key in the form of yyyy.MM.dd.dow.hh.mm.ss. Therefore, every RabbitMQ client receives the message which is bound to this queue with routing key such as "*.*.*.*.*.*.20", "2014.*.*.*.*.*.*", and so on.

You can download the metronome plugin from the **rabbitmq-metronome** repository in RabbitMQ's official Mercurial repository into your RabbitMQ development environment. Moreover, you need to make this plugin and enable the metronome plugin. Finally, run the RabbitMQ Server, and you will see that RabbitMQ executes the rabbitmq-metronome plugin. The following command lines show the process of running the metronome plugin:

```
http://hg.rabbitmq.com/rabbitmq-metronome/hg clone
http://hg.rabbitmq.com/rabbitmq-metronome/
make


mkdir -p rabbitmq-server/plugins-folder
cd rabbitmq-server/plugins-folder
ln -s rabbitmq-erlang-client
ln -s rabbitmq-metronome
scripts/rabbitmq-plugins enable rabbitmq_metronome


make run


vagrant@precise32:~$ sudo rabbitmqctl status
Status of node rabbit@precise32 ...
[{pid,844},
 {running_applications,
     [{rabbitmq_management,"RabbitMQ Management Console","3.0.4"},
      {rabbitmq_web_dispatch,"RabbitMQ Web Dispatcher","3.0.4"},
      {rabbitmq-metronome, "Embedded Rabbit Metronome", "0.01"},
      {webmachine,"webmachine","1.9.1-rmq3.0.4-git52e62bc"},
      {mochiweb,"MochiMedia Web Server","2.3.1-rmq3.0.4-gitd541e9a"},
      {rabbitmq_management_agent,"RabbitMQ Management
      Agent","3.0.4"},
      {rabbit,"RabbitMQ","3.0.4"},
      {os_mon,"CPO  CXC 138 46","2.2.7"},
      {inets,"INETS  CXC 138 49","5.7.1"},
      {xmerl,"XML parser","1.2.10"},
      {mnesia,"MNESIA  CXC 138 12","4.5"},
      {amqp_client,"RabbitMQ AMQP Client","3.0.4"},
      {sasl,"SASL  CXC 138 11","2.1.10"},
      {stdlib,"ERTS  CXC 138 10","1.17.5"},
```

```
     {kernel,"ERTS  CXC 138 10","2.14.5"}]},
 {os,{unix,linux}},
 {erlang_version,
     "Erlang R14B04 (erts-5.8.5) [source] [rq:1] [async-threads:30]
     [kernel-poll:true]\n"},
 {memory,
     [{total,16283792},
      {connection_procs,2728},
      {queue_procs,25080},
      {plugins,48952},
      {other_proc,4756524},
      {mnesia,31508},
      {mgmt_db,25444},
      {msg_index,11208},
      {other_ets,521060},
      {binary,2784},
      {code,9136933},
      {atom,1027009},
      {other_system,694562}]},
 {vm_memory_high_watermark,0.4},
 {vm_memory_limit,154828800},
 {disk_free_limit,1000000000},
 {disk_free,77275533312},
 {file_descriptors,
     [{total_limit,924},{total_used,5},{sockets_limit,829},
     {sockets_used,1}]},
 {processes,[{limit,1048576},{used,190}]},
 {run_queue,0},
 {uptime,206}]
...done.
```

Now that we have seen that our custom plugin rabbitmq-metronome worked on the RabbitMQ Server, let's move onto the details of this plugin and its codes.

Firstly, we should look over each code in rabbitmq-metronome with the following table:

Filename	Description
`rabbitmq_metronome.app.src`	This file simply defines the dependencies and the module properties such as its name, its version, and so on.
`rabbitmq_metronome.erl`	This file presents the Erlang "application" behavior and starts and stops the plugin with the related Erlang VM.
`rabbitmq_metronome_sup.erl`	This file presents the Erlang "supervisor" behavior that monitors the worker process and restarts it if it crashes.
`rabbitmq_metronome_worker.erl`	This file is the core of the plugin. All of the work is done by this code. In metronome plugin, this code connects to the RabbitMQ Server and creates a task that will be triggered every second.
`rabbitmq_metronome_tests.erl`	This file represents the tests of the plugin. You can run the tests with the following command line: **`make test`**

After talking about the overlook of the codes inside rabbitmq-metronome, we will now go into the details of the important codes, starting with `rabbitmq_metronome.app.src`:

```
% Filename: rabbitmq_metronome.app.src
{application, rabbitmq_metronome,
   [{description, "Embedded Rabbit Metronome"},
    {vsn, "0.01"},
    {modules, []},
    {registered, []},
    {mod, {rabbit_metronome, []}},
    {env, []},
    {applications, [kernel, stdlib, rabbit, amqp_client]}]}.
```

As we can see, this code simply defines the application name, version, its modules, environment variables, and its dependencies. Every module should have this kind of parameters to describe the module.

The following code shows the main functionality code called `rabbit_metronome_worker`. First we will look at the code, and then we'll discuss the code in detail:

```
%% Filename: rabbit_metronome_worker.erl
%% Copyright (c) 2007-2013 GoPivotal, Inc.
%% You may use this code for any purpose.

-module(rabbit_metronome_worker).
-behaviour(gen_server).

-export([start_link/0]).

-export([init/1, handle_call/3, handle_cast/2, handle_info/2,
         terminate/2, code_change/3]).

-export([fire/0]).

-include_lib("amqp_client/include/amqp_client.hrl").

-record(state, {channel}).

-define(RKFormat,
        "~4.10.0B.~2.10.0B.~2.10.0B.~1.10.0B.~2.10.0B.~2.10.0B.~2.10.
0B").

start_link() ->
    gen_server:start_link({global, ?MODULE}, ?MODULE, [], []).

%---------------------------
% Gen Server Implementation
% --------------------------

init([]) ->
    {ok, Connection} =
    amqp_connection:start(#amqp_params_direct{}),
    {ok, Channel} = amqp_connection:open_channel(Connection),
    amqp_channel:call(Channel, #'exchange.declare'{exchange =
    <<"metronome">>,type = <<"topic">>}),
    fire(),
    {ok, #state{channel = Channel}}.

handle_call(_Msg, _From, State) ->
    {reply, unknown_command, State}.
```

```
handle_cast(fire, State = #state{channel = Channel}) ->
    Properties = #'P_basic'{content_type = <<"text/plain">>,
    delivery_mode = 1},
    {Date={Year,Month,Day},{Hour, Min,Sec}} =
    erlang:universaltime(),
    DayOfWeek = calendar:day_of_the_week(Date),
    RoutingKey = list_to_binary(
                   io_lib:format(?RKFormat, [Year, Month, Day,
                                 DayOfWeek, Hour, Min, Sec])),
    Message = RoutingKey,
    BasicPublish = #'basic.publish'{exchange = <<"metronome">>,
                                    routing_key = RoutingKey},
    Content = #amqp_msg{props = Properties, payload = Message},
    amqp_channel:call(Channel, BasicPublish, Content),
    timer:apply_after(1000, ?MODULE, fire, []),
    {noreply, State};

handle_cast(_, State) ->
    {noreply,State}.

handle_info(_Info, State) ->
    {noreply, State}.

terminate(_, #state{channel = Channel}) ->
    amqp_channel:call(Channel, #'channel.close'{}),
    ok.

code_change(_OldVsn, State, _Extra) ->
    {ok, State}.

%---------------------------

fire() ->
    gen_server:cast({global, ?MODULE}, fire).
```

The preceding code simply performs the opening connection on initializing the server connection, and then, in every second, code sends message to the queue with routing name equals to the date and time of the message sent. If the connection to RabbitMQ Server is closed, then module's connection to the RabbitMQ Server also terminates.

Connection to the server is opened in the `init()` function. The functions with their names starting with "handle" are to be called in every RabbitMQ Server invocation. Therefore, we need to implement our message sending code into these functions. Finally, we should terminate our connection from the `terminate()` function.

The following code shows the supervisor code of the plugin. As said earlier, the supervisor module monitors the functionality of the worker. First we'll look at the source code, and then we'll dive into the code details:

```
%% Copyright (c) 2007-2013 GoPivotal, Inc.
%% You may use this code for any purpose.

-module(rabbit_metronome_sup).

-behaviour(supervisor).

-export([start_link/0, init/1]).

start_link() ->
    supervisor:start_link({local, ?MODULE}, ?MODULE, _Arg = []).

init([]) ->
    {ok, {{one_for_one, 3, 10},
           [{rabbit_metronome_worker,
             {rabbit_metronome_worker, start_link, []},
             permanent,
             10000,
             worker,
             [rabbit_metronome_worker]}
           ]}}.
```

Supervisor code just monitors the worker. Supervisor uses the `start_link()` function. The following code describes starting and stopping the `rabbitmq-metronome` plugin. RabbitMQ Server calls the `start()` function when RabbitMQ enables the plugin, and it calls the `stop()` function when RabbitMQ disables the plugin:

```
%% Copyright (c) 2007-2013 GoPivotal, Inc.
%% You may use this code for any purpose.

-module(rabbit_metronome).

-behaviour(application).

-export([start/2, stop/1]).

start(normal, []) ->
    rabbit_metronome_sup:start_link().

stop(_State) ->
    ok.
```

Summary

RabbitMQ plugins are a great way to extend the functionalities of the RabbitMQ server. Default plugins give us a chance to extend RabbitMQ in different ways such as supporting another protocol and monitoring and managing the RabbitMQ easily.

RabbitMQ is written in the Erlang runtime environment. We have to write our plugins in Erlang programming language. Thus, we talked about the basics of the Erlang where we gave lots of codes. Moreover, we programmed a custom plugin that is published by the RabbitMQ contributors called rabbitmq-metronome. It showed us how to design and develop our own plugins onto the RabbitMQ Server.

We'll now talk about the management of RabbitMQ using different types of techniques.

6
Managing Your RabbitMQ Server

After talking about the details of the plugins and plugin development, we are now ready to look into the management of the RabbitMQ server. To get the best out of RabbitMQ, we need to manage it effectively. RabbitMQ provides support for the following:

- Adding, updating, and showing users, virtual hosts, and permissions
- Declaring, listing, and deleting exchanges, queues, and bindings
- Sending and receiving messages
- Monitoring the queue length, message rates globally and per channel, data rates per connection, and so forth
- Exporting/importing object definitions to JSON
- Forcing close connections
- Purging queues

We can manage the RabbitMQ server using command-line tool called `rabbitmqctl` using a plugin called *Management Plugin*, that is provided by default from RabbitMQ and accessing RabbitMQ using the REST APIs. Therefore, our chapter is designed with the following topics:

- Management via a command line
- Management via a web plugin
- Management via a REST API

Management via a command line

The most powerful tool for managing RabbitMQ is **rabbitmqctl**, which is a command-line application that comes with the default RabbitMQ server installation bundle. Using the RabbitMQ control tool is really simple; you just need to run the tool with its parameters.

Cluster commands

In the following table, we present cluster commands:

Parameter	Description
`join_cluster {clusternode} [--ram]`	This joins the specified cluster to the main node. If a `ram` attribute is provided, RabbitMQ joins the cluster as a RAM node. For example: `rabbitmqctl join_cluster newcl@local --ram`
`forget_cluster_node [--offline]`	This removes the cluster node remotely. If an offline parameter is specified, RabbitMQ enables node removal from an offline node.
`change_cluster_node_type {disc | ram}`	This changes the type of cluster node to disc or RAM.
`cluster_status`	This displays the status of the cluster.
`update_cluster_nodes {clusternode}`	This updates all the clusters with the latest information. If the `clusternode` parameter is given, information of the specific cluster is updated.

User commands

In the following table, we present user commands:

Parameter	Description
add_user <username> <password>	This adds a new user with a given username and password.
delete_user <username>	This deletes the user that is specified by the username.
change_password <username> <password>	This changes the password with the new password of the user, specified by the username.
clear_password <username>	This clears the password of the user, specified by the username.
set_user_tags <username> <tag>	This sets the new tags to the user that are specified by the username.
list_users	This lists all the users within the RabbitMQ broker.

Virtual host and permission commands

In the following table, we present virtual host and permission commands:

Parameter	Description
add_vhost <vhostpath>	This adds the new virtual host to the RabbitMQ with the given name.
delete_vhost <vhostpath>	This deletes the virtual host that is specified by the given name.
list_vhosts [<vhostinfoitem> ...]	This lists all the virtual hosts within the RabbitMQ broker. The vhostinfoitem parameter specifies the information that is listed with the virtual hosts.
set_permissions [-p <vhostpath>] <user> <conf> <write> <read>	This sets the permission for the given user with the specified permissions, such as write and read. If vhostpath is specified, the permission of user is set in the specified virtual host.
clear_permissions [-p <vhostpath>] <username>	This clears all the permissions for the specified user. If vhostpath is specified, permission of the user is cleared in the specified virtual host.

Parameter	Description
`list_permissions [-p <vhostpath>]`	This lists all the permissions. If `vhostpath` is specified, permissions are listed in the specified virtual host.
`list_user_permissions [-p <vhostpath>] <username>`	This lists the permissions of the given user. If `vhostpath` is specified, permission of the user is listed in the specified virtual host.

Miscellaneous commands

In the following table, we present miscellaneous commands:

Parameter	Description
`set_parameter [-p vhostpath] {component_name} {name} {value}`	This sets a parameter specified by `component_name`, `name` and `value`. If `vhostpath` is specified, the parameter's setting is only effective for the specified virtual host.
`clear_parameter [-p vhostpath] {component_name} {key}`	This clears the parameter specified by `component_name` and the `key`. If `vhostpath` is specified, the parameters are cleaned, only effective for the specified virtual host.
`list_parameters [-p vhostpath]`	This lists all the parameters. If `vhostpath` is specified, parameters are listed only for the specified virtual host.
`set_policy [-p vhostpath] [--priority priority] [--apply-to apply-to] {name} {pattern} {definition}`	This sets the policy by covering the queues that are specified by the `pattern` parameter.
`clear_policy [-p vhostpath] {name}`	This clears the policy. If `vhostpath` is specified, policies are cleared for the specified virtual host.
`list_policies [-p vhostpath]`	This lists the policies. If `vhostpath` is specified, policies are listed for the specified virtual host.
`list_queues [-p <vhostpath>] [<queueinfoitem> ...]`	This lists the queues. If `vhostpath` is specified, queues are listed for the specified virtual host.

Parameter	Description
list_exchanges [-p <vhostpath>] [<exchangeinfoitem> ...]	This lists the exchanges. If vhostpath is specified, exchanges are listed for the specified virtual host.
list_bindings [-p <vhostpath>] [<bindinginfoitem> ...]	This lists the bindings. If vhostpath is specified, bindings are listed for the specified virtual host.
list_connections [<connectioninfoitem> ...]	This lists the connections.
list_channels [<channelinfoitem> ...]	This lists the channels.
list_consumers [-p <vhostpath>]	This lists the consumers. If vhostpath is specified, consumers are listed for the specified virtual host.
status	This displays the current status of the RabbitMQ broker.
environment	This displays the environment variables in the application environment.
report	This generates a report that contains the status of the system. For example: rabbitmqctl report > report.txt
eval <expr>	This executes an arbitrary Erlang expression.
close_connection <connectionpid> <explanation>	This closes the connection that is associated with the Erlang process ID connectionpid.
trace_on [-p <vhost>]	This starts the tracing. If vhost is specified, tracing is enabled only on the specified virtual host
trace_off [-p <vhost>]	This ends the tracing. If vhost is specified, tracing is disabled only on the specified virtual host.
set_vm_memory_high_watermark <fraction>	This sets the new memory threshold. rabbitmqctl set_vm_memory_high_watermark 0.4

As shown earlier, the table that lists all the parameters, we provide the parameters to `rabbitmqctl`, and then we execute the command as the screenshots showed.

The `rabbitmqctl` command is really enough for managing all the parts of the RabbitMQ servers. Therefore, it is widely used in the RabbitMQ users.

Management via a web plugin

After talking the details of the management of RabbitMQ using a command-line tool, we are now ready to talk about the management plugin. **Management plugin** is simply a web application that is written in Erlang. You can monitor and control RabbitMQ using the management web interface. Management plugin is provided as default by the RabbitMQ installation; however, you need to enable the management plugin to use it by performing the following steps:

1. Enable the management plugin with the help of the `rabbitmq-plugins` command:

   ```
   rabbitmq-plugins enable rabbitmq_management
   ```

2. You should restart the RabbitMQ Server with the following command:

   ```
   rabbitmqctl stop
   rabbitmq-server
   ```

3. Now, you are ready to open the management dashboard with the following URL:

   `http://{your-ip-address}:15672/`

4. The RabbitMQ server gives you a default username and password, that is, `guest:guest`. Note that `guest:guest` won't work for remote a RabbitMQ server later than version 3.3.

After locating the, management URL, you can see a dashboard, as shown here:

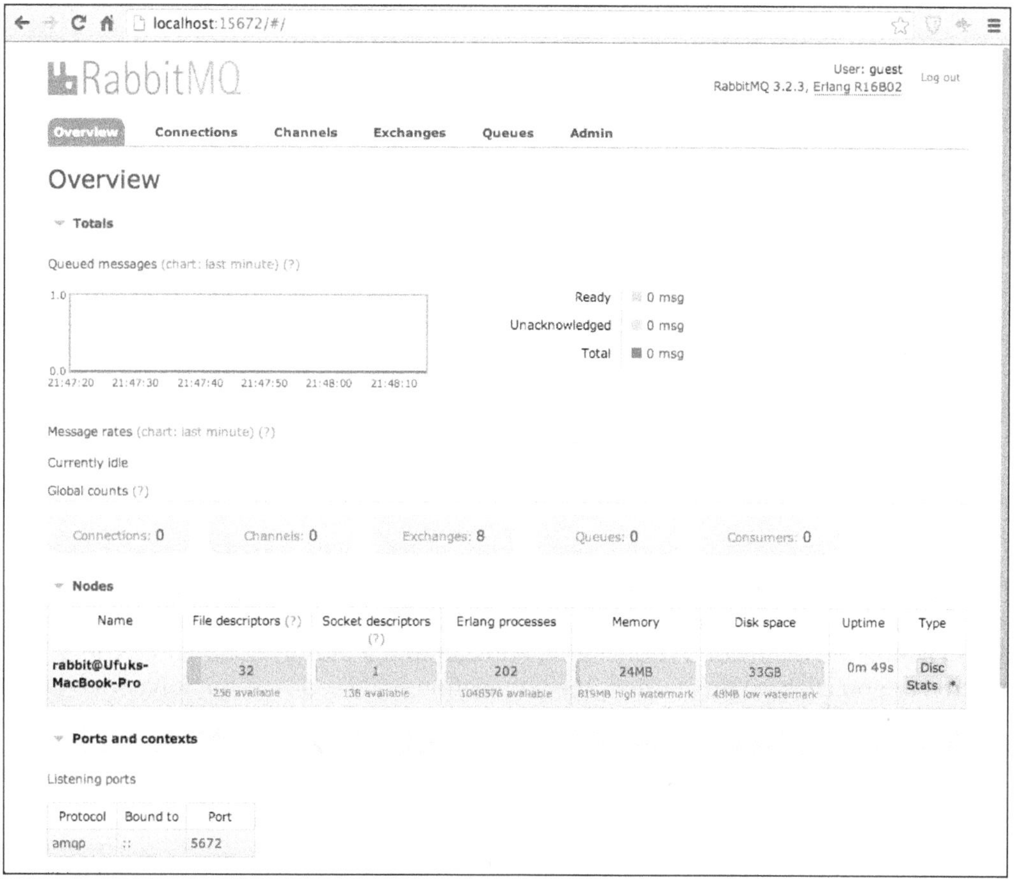

Dashboard of Management Web Interface

In the dashboard interface, you can view the RabbitMQ server statistics and information related to current connections, channels, exchanges, queues, and consumers. Additionally, the dashboard can be used for monitoring; we will cover this in *Chapter 7, Monitoring*. Moreover, you have a menu on the header side that redirects to the detailed part of each module.

After clicking on the **Connections** tab on the menu, you can see the **Connections** module in the following image:

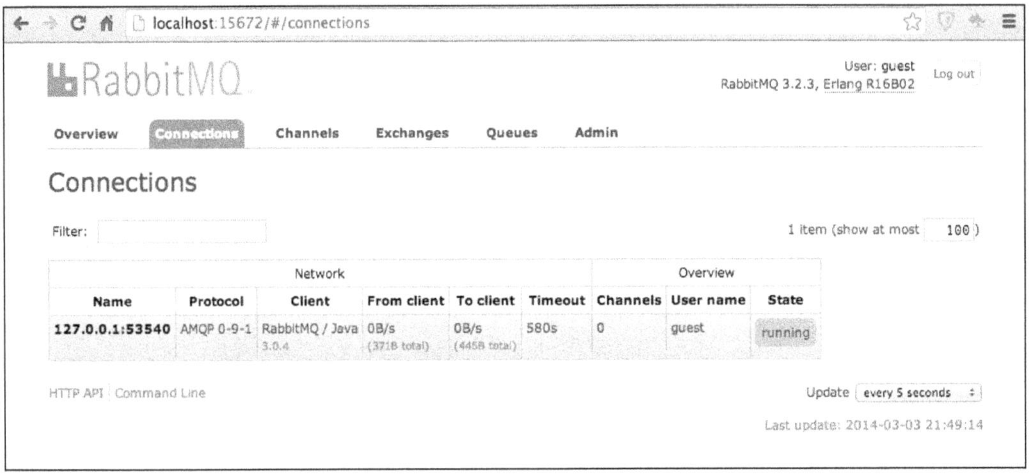

Connections

You can find the related information for each of the connections. Moreover, you are allowed to close the connection with the help of the **Connections** web page. After clicking on the related connection, you can find a button titled **Force Close**. Whenever you click on this button, the connection will be closed forcibly.

The following image simply describes the **Channels** tab and its web page:

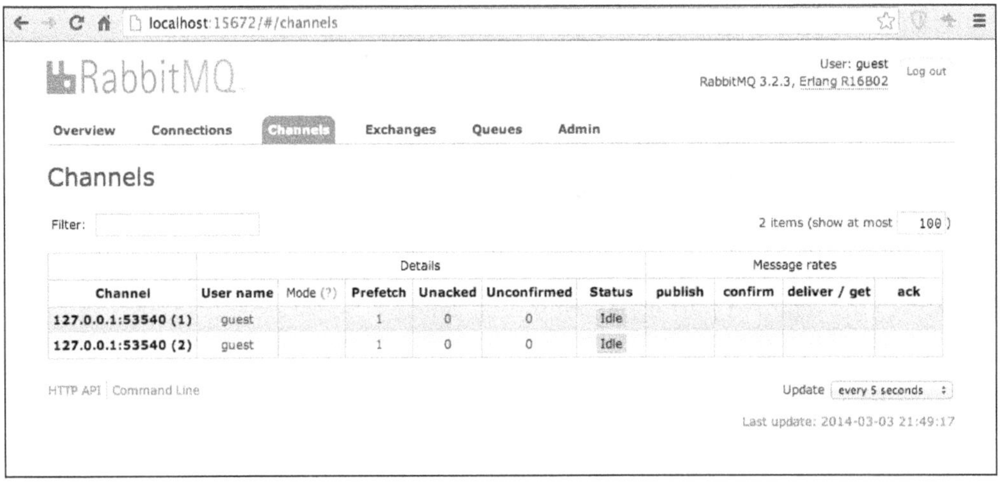

Channels

In the **Channels** web page, you can find the related information about each channel. Each channel is monitored with its message rates, connections, and so on in this web page.

The following image shows the **Exchanges** module and its web page in the Management plugin:

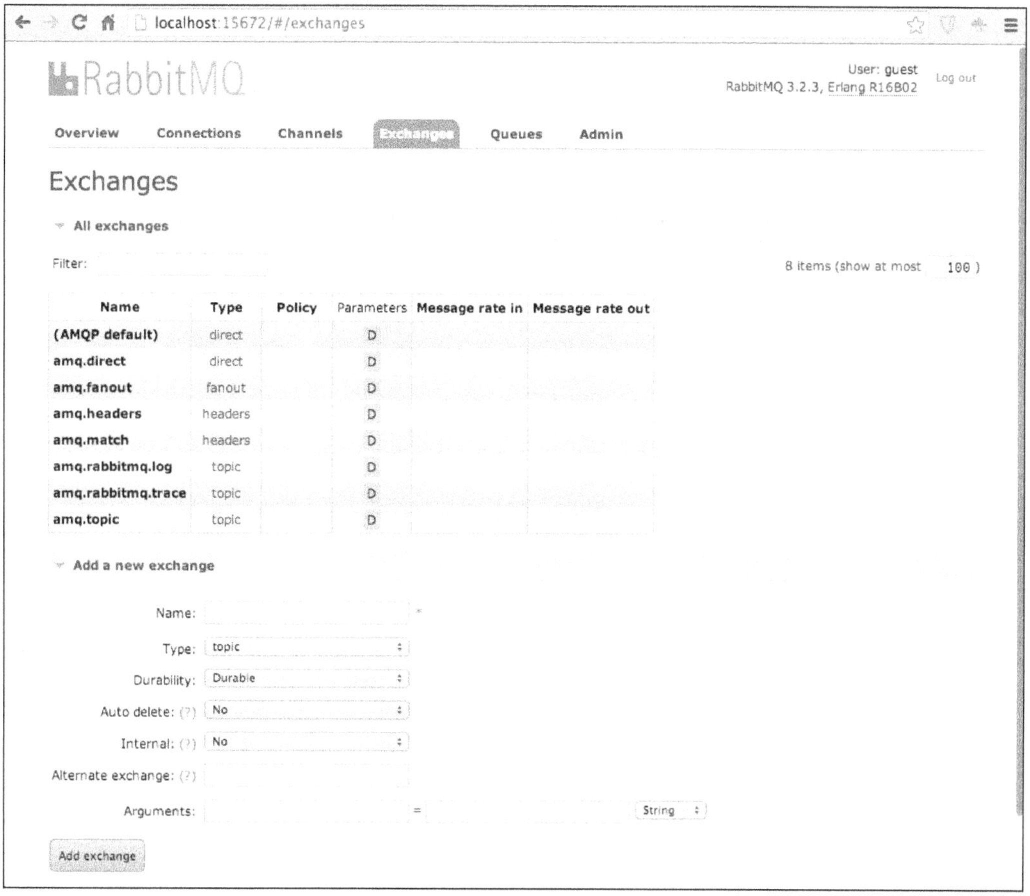

Exchanges

In the **Exchanges** web page, we can monitor all of the exchanges in the current RabbitMQ server with its related information. Moreover, if you click on each exchange, you'll get detailed information about the clicked exchange item. You can publish and delete a message through a selected exchange. Furthermore, you can delete the selected exchange as well. In the **Exchanges** web page, you are allowed to add a new exchange.

The following image shows the **Queues** web page:

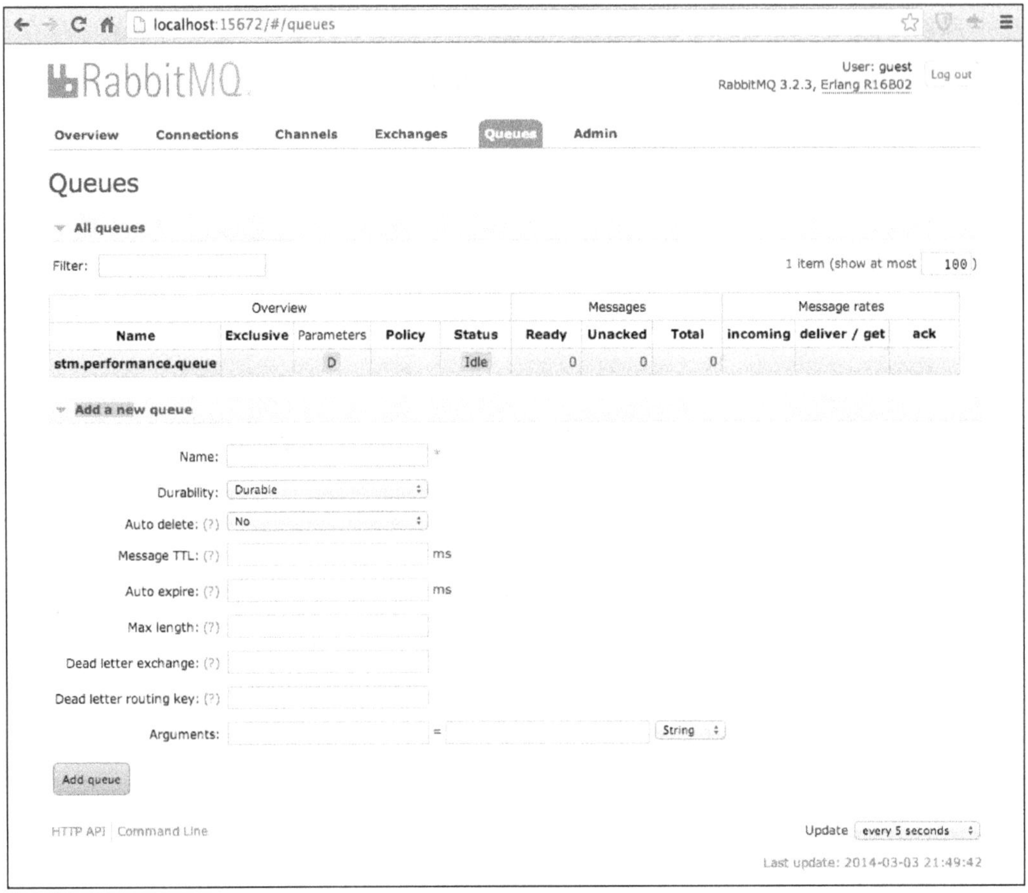

Queues

In the **Queues** web page, you can find the list of the queues and their related information. Moreover, you can add a new queue with the **Add a new queue** option and fill in the details in the required tabs. If you click on one of the queues in this web page, you can find the detailed information about the selected queue. Furthermore, you can add a new routing key for binding, publishing, and getting a message from the queue.

Let's talk about the last item of our Management plugins called the **Users** web page, which is as shown in the following image:

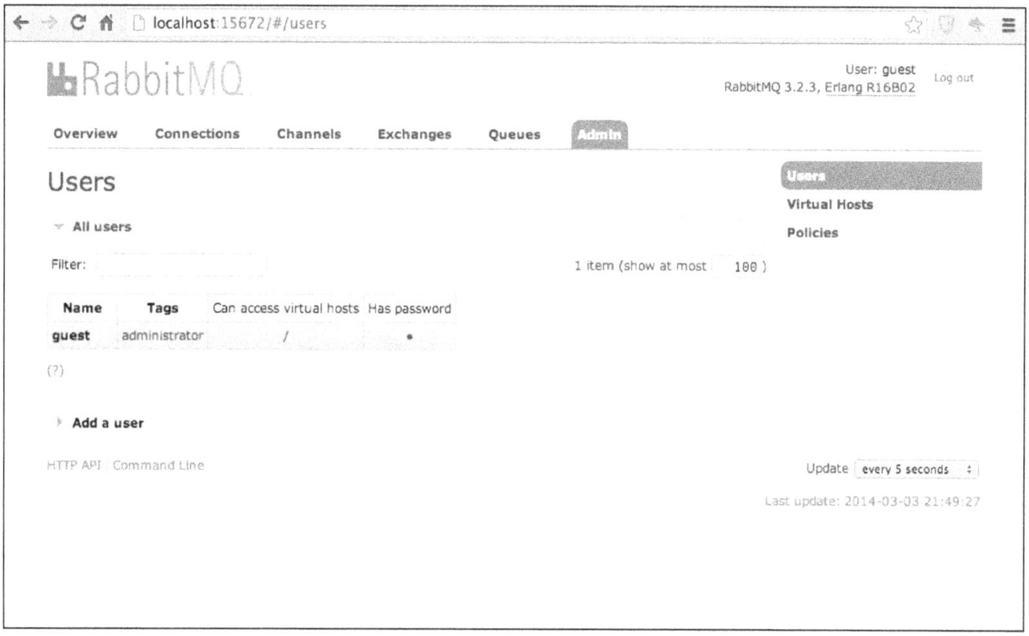

Users

Our last web page is the admin part that is accessible for only administrators. You can see the details of each user and add a new user with the help of the **Admin** web page. Additionally, the **Admin** web page gives us an opportunity to control and manage virtual hosts and policies as well.

Management via a REST API

Our last choice of management and controlling the RabbitMQ is using the **REST API**. RabbitMQ supports REST API to get lots of information from the RabbitMQ server and add, edit, and delete some parameters and properties on it.

As REST services rely on the HTTP protocol, we can easily communicate with RabbitMQ using web pages with **AJAX**, HTTP clients on every language, and so on. We'd like to show the examples that use RabbitMQ's REST API, using the **Postman** that is a REST client for Google Chrome. Postman is a free extension on Google Chrome, and you can add the Postman using the extension market of Google Chrome.

Before diving into the REST APIs, we'd like to talk about the authentication issue and return of the REST API after solving the issue. REST API of the RabbitMQ uses basic authentication and returns only **JSON** format. Therefore, we should configure our custom monitoring and managing tool with respect to these authentication and resource types. Lastly, RabbitMQ uses `55672` as a default port for the REST API port.

With the REST API, we can access the overview information about the RabbitMQ server. You should provide a username and password for basic authentication and just add the related URL for an overview of the REST API. Now, you are ready to send the request using the **Send** button, ah shown in the following image:

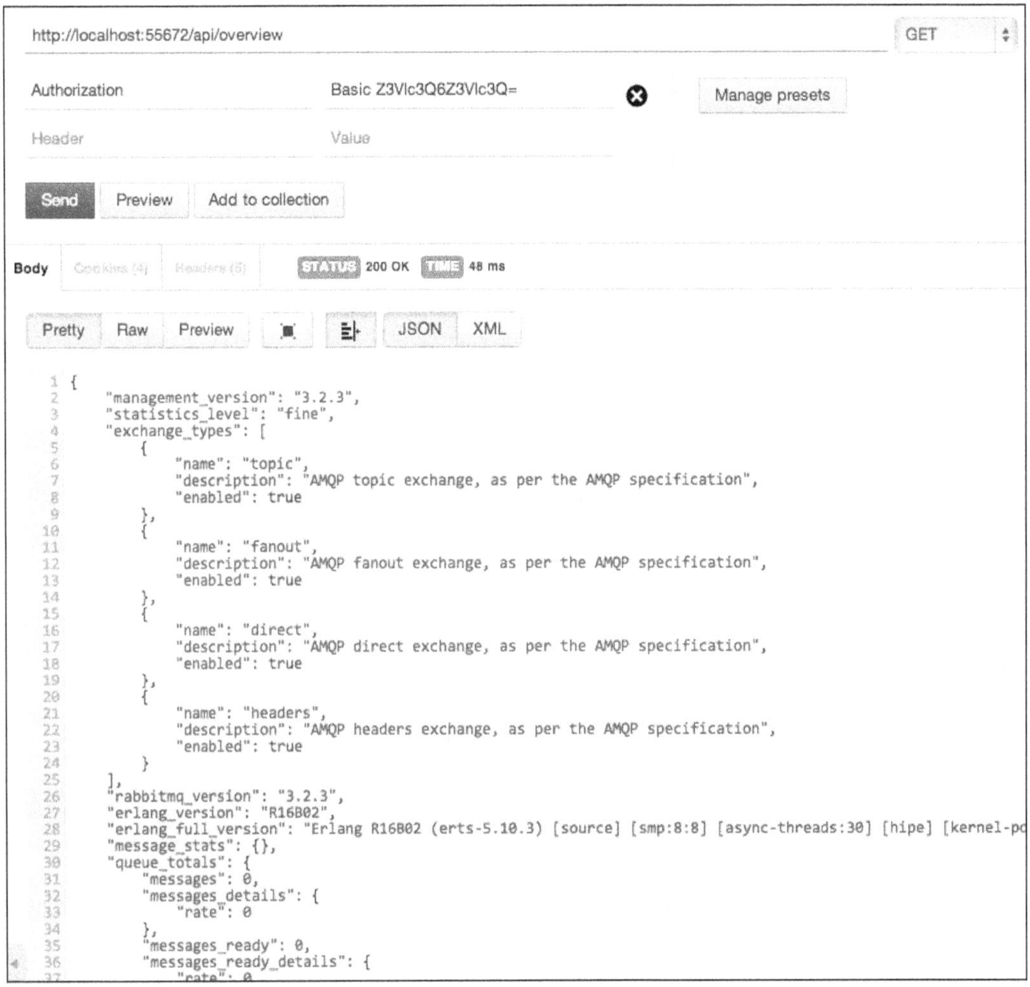

Overview Request

As we examined the screenshot of the overview result of REST API, we easily found the information and its statistics for RabbitMQ. We also found a similar view using the dashboard of the management web page.

Let's now move on to the queues and their details with the following image:

Queues Request

The queues service simply returns the list of all the queues, their information, and statistics. We can use these statistics to monitor our queues.

The following screenshot describes the connections service and type is JSON. The connections service simply returns the statistics and information about the current connections, which are established on the RabbitMQ server:

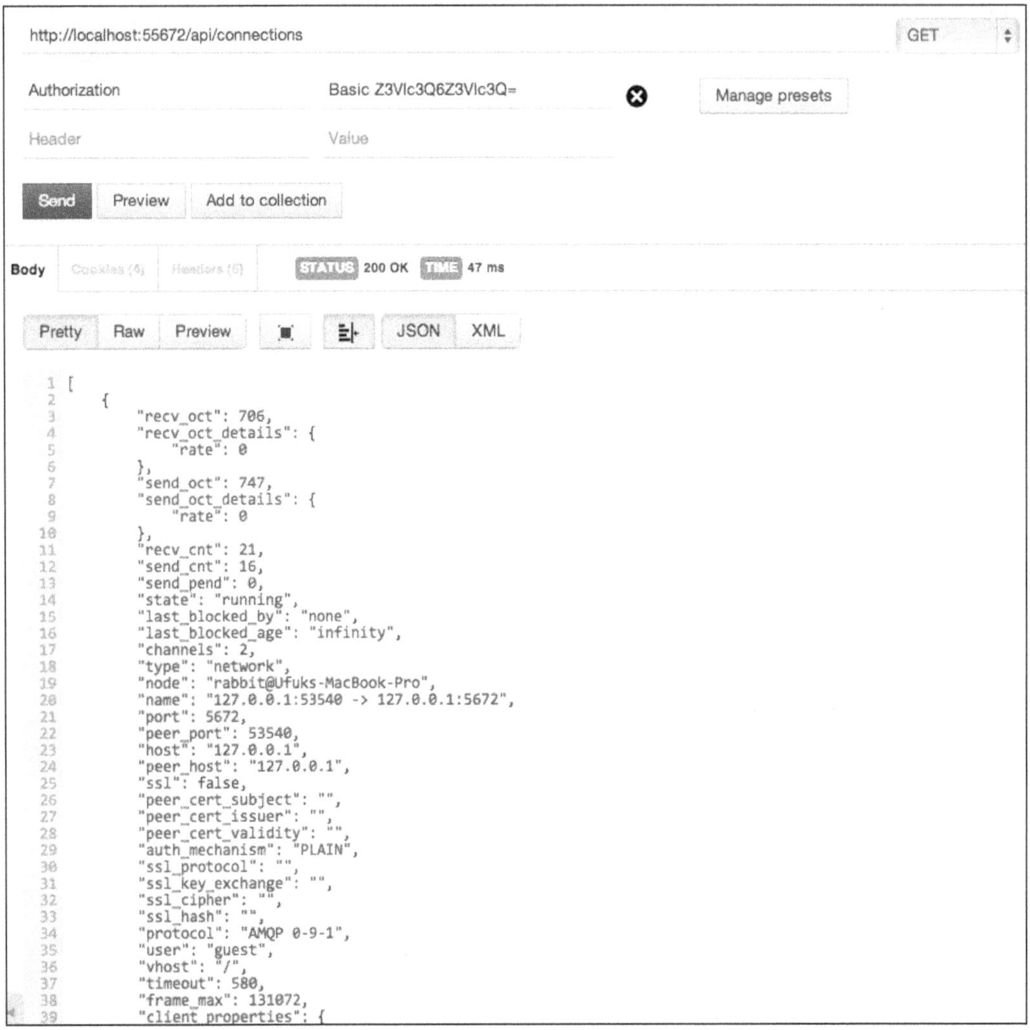

http://localhost:55672/api/connections
GET
Authorization
Basic Z3Vlc3Q6Z3Vlc3Q=
Manage presets
Header
Value
Send
Preview
Add to collection
Body
STATUS 200 OK
TIME 47 ms
Pretty
Raw
Preview
JSON
XML
[
{
"recv_oct": 706,
"recv_oct_details": {
"rate": 0
},
"send_oct": 747,
"send_oct_details": {
"rate": 0
},
"recv_cnt": 21,
"send_cnt": 16,
"send_pend": 0,
"state": "running",
"last_blocked_by": "none",
"last_blocked_age": "infinity",
"channels": 2,
"type": "network",
"node": "rabbit@Ufuks-MacBook-Pro",
"name": "127.0.0.1:53540 -> 127.0.0.1:5672",
"port": 5672,
"peer_port": 53540,
"host": "127.0.0.1",
"peer_host": "127.0.0.1",
"ssl": false,
"peer_cert_subject": "",
"peer_cert_issuer": "",
"peer_cert_validity": "",
"auth_mechanism": "PLAIN",
"ssl_protocol": "",
"ssl_key_exchange": "",
"ssl_cipher": "",
"ssl_hash": "",
"protocol": "AMQP 0-9-1",
"user": "guest",
"vhost": "/",
"timeout": 580,
"frame_max": 131072,
"client_properties": {

Connections Request

Similarly, we can control and monitor the channels using the REST API. As you can see in the following screenshot, we can fetch the information and statistics about the channels in the RabbitMQ server:

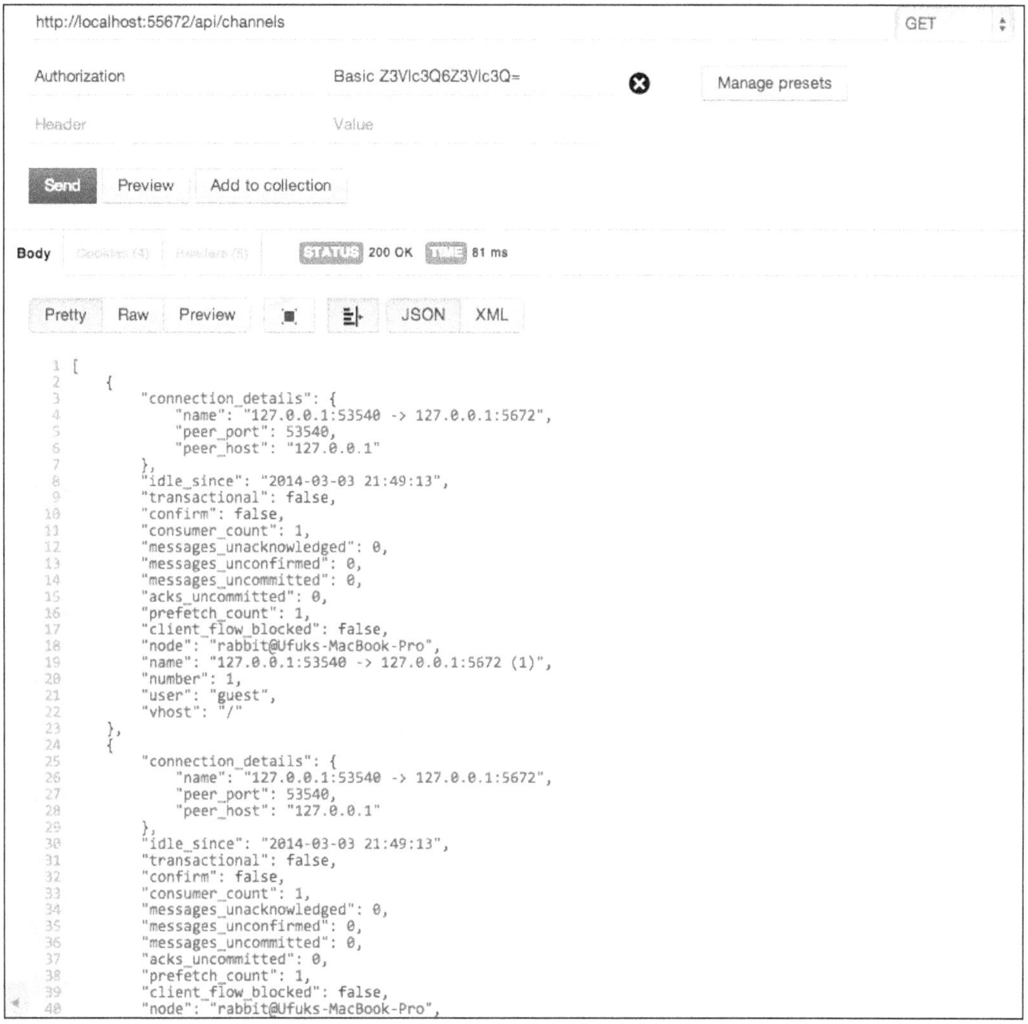

Channels Request

Statistics and information about the bindings can be easily fetched from the REST API as well.

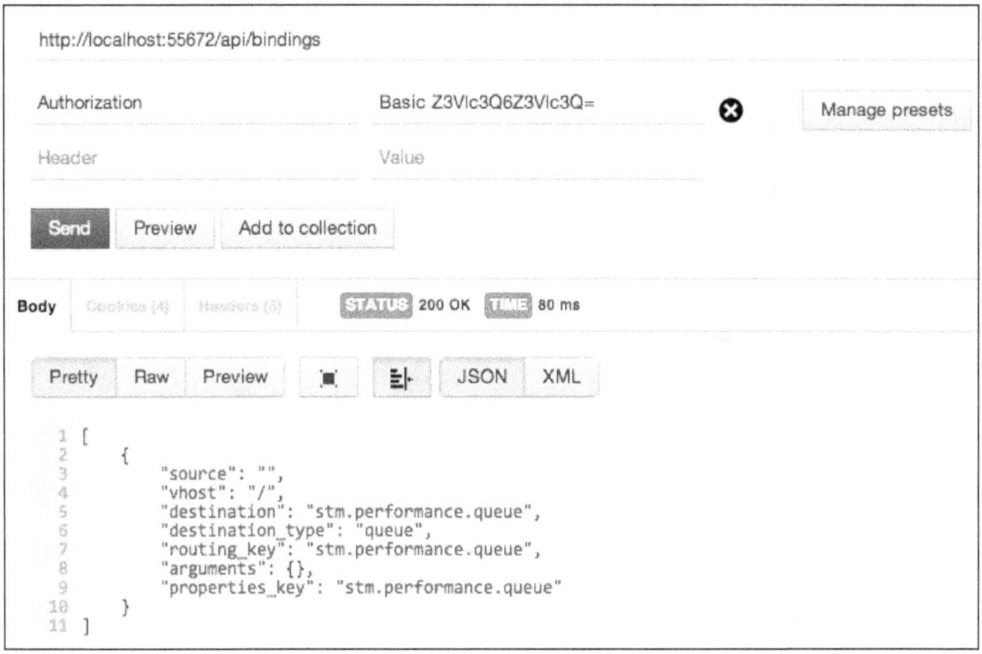

Bindings Request

We sometimes need to view the permissions of the user within the RabbitMQ server instance. With the help of RabbitMQ's REST service, we can easily fetch and show the permissions of the user as shown in the following screenshot:

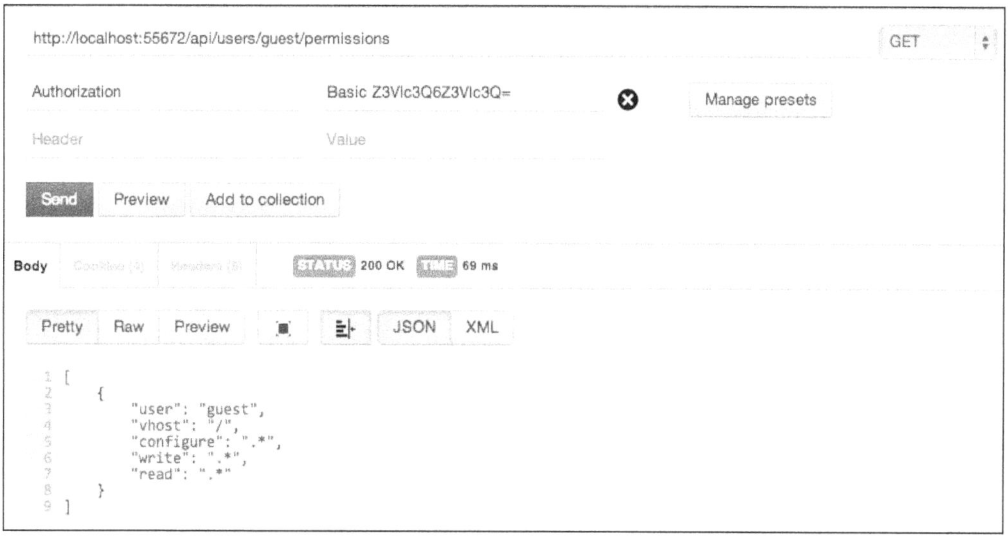

Permissions Request

As seen in the screenshots, both the services and results are in the JSON format. We can easily integrate them with our software system using the service-oriented architectures.

The RabbitMQ REST API has lots of services. Therefore, we can't show all of the services within screenshots of the each REST service. So, it is good to list all the services in a table. The following table just shows each service of the REST API of RabbitMQ with its HTTP methods and its description that explains the functionality of these parameters:

URL path	Available HTTP methods and description
`/api/overview`	HTTP Method: `GET` This returns the state and related information about the RabbitMQ Broker.
`/api/nodes/`	HTTP Method: `GET` This lists the nodes in the RabbitMQ cluster.
`/api/nodes/name`	HTTP Method: `GET` This returns the information about the node that is specified with its name.

URL path	Available HTTP methods and description
`/api/extensions/`	HTTP Method: `GET` This returns the list of extensions in the Management plugin.
`/api/definitions`	HTTP Method: `GET`, `POST` If the `GET` method is used, server definitions such as exchanges, queues, bindings, users, and virtual hosts are returned. The `POST` method is used for uploading the existing set of definitions.
`/api/connections`	HTTP Method: `GET` This returns the list of open connections.
`/api/connections/name`	HTTP Method: `GET`, `DELETE` The `GET` method is used for fetching the information of the connection that is specified with a name. The `DELETE` method is used for closing the connection.
`/api/channels`	HTTP Method: `GET` This returns the list of all open channels.
`/api/channels/channel`	HTTP Method: `GET` This returns the information of the channel that is specified with its name.
`/api/exchanges`	HTTP Method: `GET` This returns a list of the exchanges.
`/api/exchanges/vhost`	HTTP Method: `GET` This returns the list of the exchanges within the virtual host that is specified by the `vhost` parameter.
`/api/queues`	HTTP Method: `GET` This returns the list of queues.
`/api/queues/vhost`	HTTP Method: `GET` This returns the list of the queues within the virtual host that is specified by the `vhost` parameter.
`/api/bindings`	HTTP Method: `GET` This returns the list of bindings.
`/api/bindings/vhost`	HTTP Method: `GET` This returns the list of bindings within the virtual host, specified by `vhost` parameter.

URL path	Available HTTP methods and description
`/api/vhosts`	HTTP Method: `GET` This returns the list of the virtual hosts.
`/api/vhosts/name`	HTTP Method: `GET`, `PUT`, `DELETE` If the `GET` method is used, it returns the information of the virtual host that is specified by its name. The `DELETE` method is used for deleting the virtual host that is specified by its name. The `PUT` method is used for putting a virtual host.
`/api/users`	HTTP Method: `GET` This returns the list of the users.
`/api/users/name`	HTTP Method: `GET`, `PUT`, `DELETE` If the `GET` method is used, it returns the information of the users that is specified by its name. The `DELETE` method is used for deleting the user that is specified by its name. The `PUT` method is used for putting a user.
`/api/whoami`	HTTP Method: `GET` This returns the information of the current authenticated user.
`/api/permissions`	HTTP Method: `GET` This returns the list of permissions.
`/api/permissions/vhost/user`	HTTP Method: `GET`, `PUT`, `DELETE` If the `GET` method is used, it returns the information of the permissions that is specified by its virtual host and user. The `DELETE` method is used for deleting the permissions of the user. The `PUT` method is used for putting a policy.
`/api/parameters`	HTTP Method: `GET` This returns the list of the parameters.
`/api/parameters/component`	HTTP Method: `GET` This returns the list of the parameters for the provided component.
`/api/policies`	HTTP Method: `GET` This returns the list of policies.

URL path	Available HTTP methods and description
/api/policies/vhost	HTTP Method: GET This returns the list of policies within the virtual host that is specified by the vhost parameter.
/api/aliveness-test/vhost	HTTP Method: GET This makes tests for the given virtual host of the RabbitMQ server .

Summary

As we have seen, managing RabbitMQ is easy with the tools it provides. RabbitMQ gives us three opportunities to manage it using its command-line tool `rabbitmqctl`, Management plugin, and REST API. Therefore, we are comfortably managing our RabbitMQ server instances using the RabbitMQ provided structures.

In the next chapter, we will introduce the monitoring of the RabbitMQ server instances, such as monitoring the resource usage, monitoring the internal structures of RabbitMQ, and so on.

7 Monitoring

A lot of software systems among the ones we use nowadays run on our server instances. Each software system uses resources such as memory, CPU, and so on. Therefore, we should check the resource usage of each software system. We now come to the definition of monitoring. Monitoring simply means reporting or checking the software systems, their resource usage, and alerting when these levels reach a critical level so that the issue can be addressed. We need to monitor all our software systems all the time; therefore, we should monitor our RabbitMQ server too.

RabbitMQ gives us amazing tools, such as the command-line application called **rabbitmqctl** and the management web plugin. Furthermore, we also have general tools for the monitoring of server instances such as **Nagios**, **Munin**, **Zabbix**, and so on. In this chapter, we will look at each of the following monitoring tools one by one:

- The rabbitmqctl command-line application
- Management plugin
- Nagios
- Munin
- Zabbix

RabbitMQ command-line tools

RabbitMQ's powerful command-line tools have lots of skills, such as controlling, managing, and monitoring. One of the command-line tools of RabbitMQ, `rabbitmqctl`, gives us an opportunity to monitor the RabbitMQ in real-time. We use the `rabbitmqctl` tool for its monitoring functions, such as reporting on the RabbitMQ and displaying the status and specific functions of RabbitMQ.

The `report` function of the `rabbitmqctl` tool shows lots of details of RabbitMQ in realtime. The `report` function shows the sum of all monitoring results of the other functions of `rabbitmqctl`. We can find the running environment variables, configuration parameters, and cluster statuses, as shown in the next command block.

Moreover, the `report` function shows the current state of each functional structure of RabbitMQ, such as connections to RabbitMQ, channels in RabbitMQ, and so on, as shown in the following code:

```
Cluster status of node rabbit@localhost ...
[{nodes,[{disc,[rabbit@localhost]}]},
 {running_nodes,[rabbit@localhost]},
 {cluster_name,<<"rabbit@yaytas">>},
 {partitions,[]}]

Application environment of node rabbit@localhost ...
[{amqp_client,[{prefer_ipv6,false},{ssl_options,[]}]},
 {inets,[]},
 {kernel,
      [{error_logger,tty},
       {inet_default_connect_options,[{nodelay,true}]},
       {inet_dist_listen_max,25672},
       {inet_dist_listen_min,25672}]},
 {mnesia,[{dir,"/usr/local/var/lib/rabbitmq/mnesia/rabbit@
localhost"}]},
 {mochiweb,[]},
 {os_mon,
      [{start_cpu_sup,false},
       {start_disksup,false},
       {start_memsup,false},
       {start_os_sup,false}]},
 {rabbit,
      [{auth_backends,[rabbit_auth_backend_internal]},
       {auth_mechanisms,['PLAIN','AMQPLAIN']},
       {backing_queue_module,rabbit_priority_queue},
       {channel_max,0},
       {cluster_keepalive_interval,10000},
       {cluster_nodes,{[],disc}},
       {cluster_partition_handling,ignore},
       {collect_statistics,fine},
       {collect_statistics_interval,5000},
       {credit_flow_default_credit,{200,50}},
       {default_permissions,[<<".*">>,<<".*">>,<<".*">>]},
       {default_user,<<"guest">>},
       {default_user_tags,[administrator]},
```

```
        {default_vhost,<<"/">>},
        {delegate_count,16},
        {disk_free_limit,50000000},
```

On the other hand, RabbitMQ gives us another way to monitor each of the reporting items one by one. As shown in the next command block, we can monitor the running instance, configuration properties, and modules that run on the current RabbitMQ with the help of the `status` function of the `rabbitmqctl` tool:

```
Cluster status of node rabbit@localhost ...
[{nodes,[{disc,[rabbit@localhost]}]},
 {running_nodes,[rabbit@localhost]},
 {cluster_name,<<"rabbit@yaytas">>},
 {partitions,[]}]

Application environment of node rabbit@localhost ...
[{amqp_client,[{prefer_ipv6,false},{ssl_options,[]}]},
 {inets,[]},
 {kernel,
     [{error_logger,tty},
      {inet_default_connect_options,[{nodelay,true}]},
      {inet_dist_listen_max,25672},
      {inet_dist_listen_min,25672}]},
 {mnesia,[{dir,"/usr/local/var/lib/rabbitmq/mnesia/
 rabbit@localhost"}]},
 {mochiweb,[]},
 {os_mon,
     [{start_cpu_sup,false},
      {start_disksup,false},
      {start_memsup,false},
      {start_os_sup,false}]},
 {rabbit,
     [{auth_backends,[rabbit_auth_backend_internal]},
      {auth_mechanisms,['PLAIN','AMQPLAIN']},
      {backing_queue_module,rabbit_priority_queue},
      {channel_max,0},
      {cluster_keepalive_interval,10000},
      {cluster_nodes,{[],disc}},
      {cluster_partition_handling,ignore},
      {collect_statistics,fine},
      {collect_statistics_interval,5000},
      {credit_flow_default_credit,{200,50}},
      {default_permissions,[<<".*">>,<<".*">>,<<".*">>]},
      {default_user,<<"guest">>},
      {default_user_tags,[administrator]},
```

```
        {default_vhost,<<"/">>},
        {delegate_count,16},
        {disk_free_limit,50000000},

yaytas:~ yaytas$ rabbitmqctl status
Status of node rabbit@localhost ...
[{pid,9430},
 {running_applications,
     [{rabbitmq_management_visualiser,"RabbitMQ
     Visualiser","3.5.5"},
      {rabbitmq_management,"RabbitMQ Management Console","3.5.5"},
      {rabbitmq_web_dispatch,"RabbitMQ Web Dispatcher","3.5.5"},
      {webmachine,"webmachine","1.10.3-rmq3.5.5-gite9359c7"},
      {mochiweb,"MochiMedia Web Server","2.7.0-rmq3.5.5-
      git680dba8"},
      {rabbitmq_mqtt,"RabbitMQ MQTT Adapter","3.5.5"},
      {rabbitmq_stomp,"Embedded Rabbit Stomp Adapter","3.5.5"},
      {rabbitmq_management_agent,"RabbitMQ Management
      Agent","3.5.5"},
      {rabbitmq_amqp1_0,"AMQP 1.0 support for RabbitMQ","3.5.5"},
      {rabbit,"RabbitMQ","3.5.5"},
      {mnesia,"MNESIA  CXC 138 12","4.12.5"},
      {os_mon,"CPO  CXC 138 46","2.3.1"},
      {inets,"INETS  CXC 138 49","5.10.6"},
      {amqp_client,"RabbitMQ AMQP Client","3.5.5"},
      {xmerl,"XML parser","1.3.7"},
      {sasl,"SASL  CXC 138 11","2.4.1"},
      {stdlib,"ERTS  CXC 138 10","2.4"},
      {kernel,"ERTS  CXC 138 10","3.2"}]},
 {os,{unix,darwin}},
 {erlang_version,
     "Erlang/OTP 17 [erts-6.4] [source] [64-bit] [smp:8:8] [async-
     threads:64] [hipe] [kernel-poll:true]\n"},
 {memory,
     [{total,43373712},
      {connection_readers,0},
      {connection_writers,0},
      {connection_channels,0},
      {connection_other,5616},
      {queue_procs,2808},
      {queue_slave_procs,0},
      {plugins,623056},
      {other_proc,13892360},
      {mnesia,62400},
      {mgmt_db,171400},
```

```
          {msg_index,47680},
          {other_ets,1274264},
          {binary,16128},
          {code,20748464},
          {atom,711569},
          {other_system,5817967}]},
     {alarms,[]},
     {listeners,
         [{clustering,25672,"::"},
          {amqp,5672,"127.0.0.1"},
          {stomp,61613,"::"},
          {mqtt,1883,"::"}]},
     {vm_memory_high_watermark,0.4},
    {vm_memory_limit,6216758067},
     {disk_free_limit,50000000},
     {disk_free,197295443968},
     {file_descriptors,
         [{total_limit,156},{total_used,5},{sockets_limit,138},
         {sockets_used,3}]},
     {processes,[{limit,1048576},{used,200}]},
     {run_queue,0},
     {uptime,754}]
```

Starting from the `status` function, we'll look at the current RabbitMQ interactions. For instance, we need to monitor the current consumers in our RabbitMQ server instance. As the `rabbitmqctl` tool lets us list all the consumers within the server instance using the `list_consumers` function, as shown in the following command line:

```
vagrant@precise32:~$ sudo rabbitmqctl list_consumers
Listing consumers ...
stm.performance.queue <rabbit@precise32.2.312.0> amq.ctag-
PNwQde_aIWdpB2KVwTAG8A     true []
stm.performance.queue <rabbit@precise32.2.316.0> amq.ctag-P7VHTEYl-
emdaPkfsYRlXw true []
...done.
```

Moreover, we sometimes need to check the current channels on the RabbitMQ server. As shown in the following command line, the `rabbitmqctl` tool allows us to monitor the current channels on the server instance using the `list_channels` function:

```
vagrant@precise32:~$ sudo rabbitmqctl list_channels
Listing channels ...
<rabbit@precise32.2.312.0> guest 1 0
```

```
<rabbit@precise32.2.316.0> guest 1 0
...done.
```

Furthermore, we should check the current connected users. The `rabbitmqctl` tool gives us another way to list the current connected users using the `list_connections` function of RabbitMQ, as shown in the following command line:

```
vagrant@precise32:~$ sudo rabbitmqctl list_connections
Listing connections ...
guest 10.0.2.2 59144 running
...done.
```

Additionally, we have to control the current bindings on our queues in the RabbitMQ instance. To monitor bindings on the RabbitMQ queues, we can use **rabbitmqctl** to monitor the current bindings using the `list_bindings` function as shown in the following command line:

```
vagrant@precise32:~$ sudo rabbitmqctl list_bindings
Listing bindings ...
exchange stm.performance.queue queue stm.performance.queue []
...done.
```

Exchanges are the most important functions of the RabbitMQ. We can control and monitor the current exchanges. Therefore, the `rabbitmqctl` tool gives us another function to monitor the current exchanges using `list_exchanges`, as shown in the following command line:

```
vagrant@precise32:~$ sudo rabbitmqctl list_exchanges
Listing exchanges ...
direct
amq.direct direct
amq.fanout fanout
amq.headers headers
amq.match headers
amq.rabbitmq.log topic
amq.rabbitmq.trace topic
amq.topic topic
...done.
```

As we know that the queues are the main data structure of the message brokers, we need to check the current queues on the RabbitMQ server instance. The following command line simply describes the current queues on the RabbitMQ server instance using rabbitmqctl's `list_queues` function:

```
vagrant@precise32:~$ sudo rabbitmqctl list_queues
Listing queues ...
stm.performance.queue 0
...done.
```

Permissions are used for controlling the access for the different modules of the RabbitMQ server. Permissions are covered in detail in *Chapter 8, Security in RabbitMQ*. We should check the current permissions of the users on the RabbitMQ server instance. Then, we come to rabbitmqctl's function for listing permissions, that is, `list_permissions`:

```
vagrant@precise32:~$ sudo rabbitmqctl list_permissions
Listing permissions in vhost "/" ...
guest .* .* .*
monit .* .* .*
monitor .* .* .*
...done.
```

With the `list_permissions` function, we are able to list the permissions of the users. Additionally, we need to list the current users with their tags that are related with their roles. The `rabbitmqctl` tool gives us another function to list users, that is, the `list_users` function, as shown in the following command line:

```
vagrant@precise32:~$ sudo rabbitmqctl list_users
Listing users ...
guest [administrator]
monit [administrator]
monitor [monitoring]
...done.
```

Finally, we need to list the virtual hosts on the RabbitMQ server instance. Our powerful `rabbitmqctl` tool gives us another great function called `list_vhosts` to list all the virtual hosts as shown in the following command line:

```
vagrant@precise32:~$ sudo rabbitmqctl list_vhosts
Listing vhosts ...
/
...done.
```

As we looked into the details of each of the function of RabbitMQ's powerful control and monitoring command line called `rabbitmqctl`, we are able to monitor each statistical information and details of the RabbitMQ with real-time support.

Web plugins

As we know from *Chapter 6, Managing Your RabbitMQ Server*, the RabbitMQ management plugin provides an HTTP-based API for management and monitoring of the RabbitMQ server with a browser-based web user interface and command-line tool called **rabbitmqadmin**. The management plugin has a lot of monitoring features, as listed here:

- Monitors queue length and message rates
- Monitors Erlang processes
- Monitors memory use
- Monitor connections and exchanges
- Monitor users and their permissions

The following screenshot describes the dashboard screen of the RabbitMQ management web UI, where we can see number of connections, channels, exchanges, queues, and current consumers:

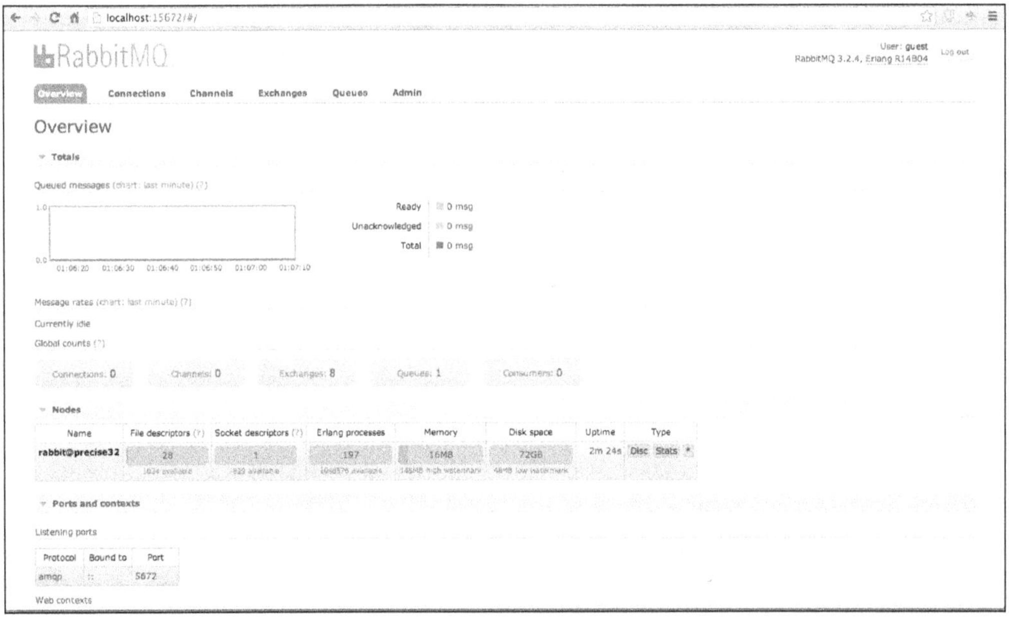

RabbitMQ Management Web Interface

Nagios

Before talking about the details of the **Nagios** RabbitMQ plugin and monitoring with the help of Nagios, we'd like to dive into the basics of Nagios. Nagios is simply defined as a powerful monitoring system that enables organizations to identify and resolve IT problems before affecting the business processes. It is a powerful tool that provides monitoring, alerting, reporting, maintenance, and planning of server.

Nagios has lots of functionalities as covered in the following list:

- Monitoring your entire infrastructure
- Responding to the issues for the limits and problems
- Automatically fixing the problems when they are detected
- Coordinating the technical team responses

Nagios is an extendable tool with its plugins. RabbitMQ can be bridged with the Nagios plugin. The Nagios-RabbitMQ plugin is developed by James Casey. It is an open source project and is published on Github. We need the RabbitMQ management plugin to use this open source project in Nagios.

First, we will download the source code using the `git scm` tool. Then, we will copy all of the scripts to the Nagios plugin directory, as shown in the following command lines:

```
git clone https://github.com/jamesc/nagios-plugins-rabbitmq.git
cd nagios-plugins-rabbitmq/scripts
cp * /usr/lib/nagios/plugins/
```

As the Nagios-RabbitMQ plugin is developed in **Perl**, it requires a few Perl libraries such as the Nagios plugin library and JSON library to communicate with RabbitMQ, as shown in the following command line:

```
sudo apt-get install libnagios-plugin-perl libjson-perl
```

Note that the installation command can be different for different operating systems. Now, we are ready to execute the scripts of the plugin. We should give the `hostname`, `port`, `username`, and `password` as parameters to the scripts. The following image shows how the overview script of RabbitMQ is executed:

Nagios RabbitMQ Command

After testing each command, we are now ready to integrate our commands with RabbitMQ. We should define our commands in the commands `configuration` file that is available in the directory of Nagios `configuration` files. We should define the place of the command line and its parameters as given in the format of `"$ARG1$"`. For our plugin, we have to provide a `hostname`, `port`, `username`, `password`, and other details with the parameter:

```
# vi /etc/nagios/commands.cfg
define command {
   command_name check_rabbitmq_server
   command_line $USER1$/nagios-plugins-
   rabbitmq/scripts/check_rabbitmq_server -H $ARG1$ --port=$ARG2$ -u
   $ARG3$ -p $ARG4$
}
```

Moreover, we should define the service for Nagios. We can provide service-related information and command-related parameters in the service definition file that is saved in the `services` directory of the Nagios `configuration` files:

```
# cat rabbitmq-service.cfg
define service {
    use                     generic-service
    host_name               dev-db
    service_description     RabbitMQ
```

```
    contacts                    prodalert
    check_command               check_rabbitmq_server!dev-
db!15672!guest!MySecretPassword
}
```

Finally, we completed the integration with RabbitMQ and Nagios with the help of a plugin. After integration, if you check the Nagios web interface, you may not see the information related to RabbitMQ. After some time, Nagios starts monitoring the RabbitMQ using the commands that you specified, as shown in the following screenshot:

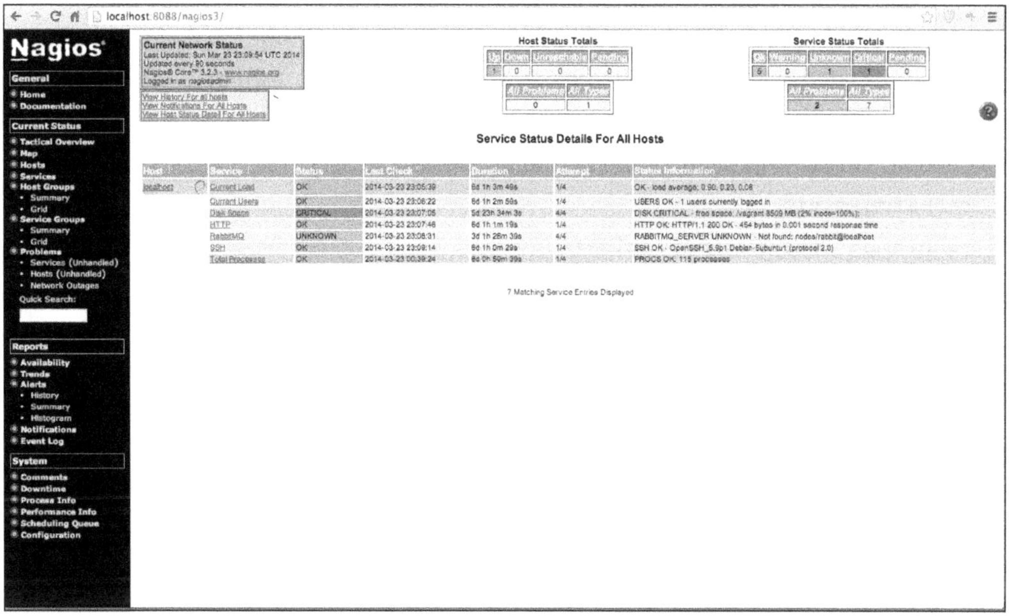

Nagios web interface

Nagios is a widely market-accepted and powerful monitoring tool for the systems and monitoring RabbitMQ is really simple using Nagios. In addition to `check_rabbitmq_server`, there are other checks supported by this plugin:

- `check_rabbitmq_aliveness`
- `check_rabbitmq_objects`
- `check_rabbitmq_overview`
- `check_rabbitmq_queue`
- `check_rabbitmq_watermark`

Munin

Munin is another powerful tool for monitoring the systems. Munin is simply defined as a networked resource monitoring tool that can help analyze resource trends, and it is a detector of the performance problems according to the Munin official web site. Munin is easily expandable with its plugins.

We can monitor the resource usage of RabbitMQ using Munin with the help of **Munin-RabbitMQ** plugin. Ask Solem Hoel who is the employee of the **Pivotal** that is the main contributor is the main contributor for Munin-RabbitMQ plugin. He works for Pivotal which is main contributor to RabbitMQ. The plugin is open source and is published on GitHub. Before the integration of the RabbitMQ and Munin, we need to download all the source code using the `git scm` tool and copy all of the RabbitMQ-related files to the directory of Munin plugins, as shown in the following command line:

```
git clone https://github.com/ask/rabbitmq-munin.git
cd rabbitmq-munin
cp rabbitmq* /etc/munin/plugins/
```

After copying all of the files into the Munin `plugin` folder, we will introduce the plugin to Munin with the help of the plugin `configuration` file that is located in the Munin `configuration` folder. Each file has to be introduced to the `configuration` file and a user has to to be provided:

```
[rabbitmq_connections]
user root

[rabbitmq_consumers]
user root

[rabbitmq_messages]
user root

[rabbitmq_messages_unacknowledged]
user root

[rabbitmq_messages_uncommitted]
user root

[rabbitmq_queue_memory]
user root
```

After the initialization, we will be able to monitor our RabbitMQ resource usage with the help of Munin dashboard, as shown in the following image:

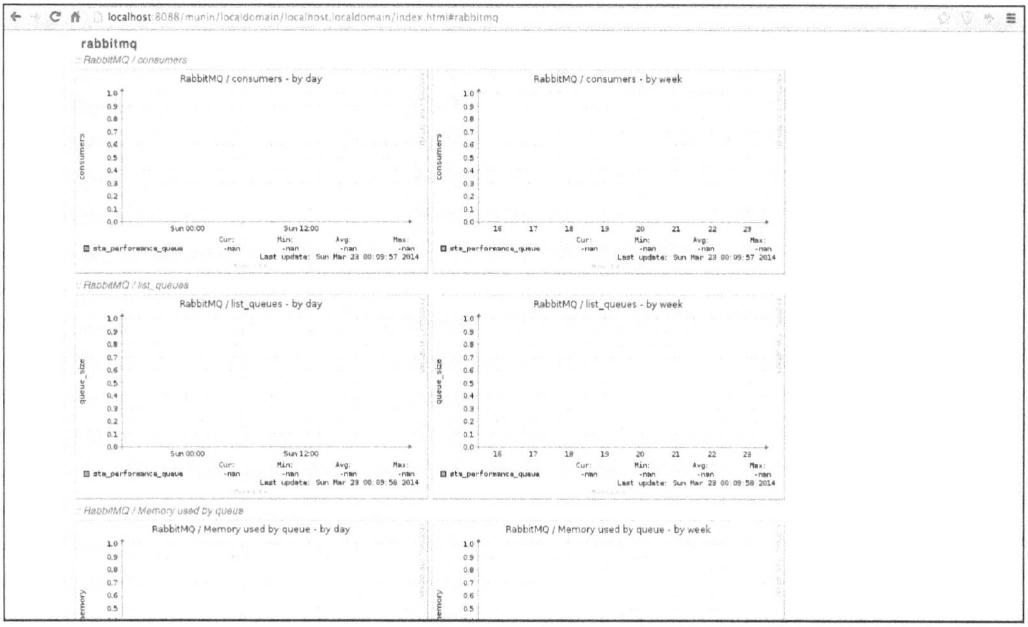

Munin web interface

Zabbix

Another powerful tool for monitoring is called **Zabbix**. Zabbix is an open source project and is released under the GPL license. Therefore, it is free of charge for both commercial and noncommercial use. According to the Zabbix official website, Zabbix is the ultimate enterprise-level software designed for monitoring availability and performance of IT infrastructure. Zabbix is developed on the **LAMP** platform (**Linux**, **Apache**, **MySQL**, and **PHP**).

Zabbix provides following functionalities:

- Collecting data from many kinds of sources
- Detecting problems
- Visualizing the collected data meaningfully
- Notifying the related users about the created events
- Supporting distributed monitoring

Let's take a look at the following screenshot:

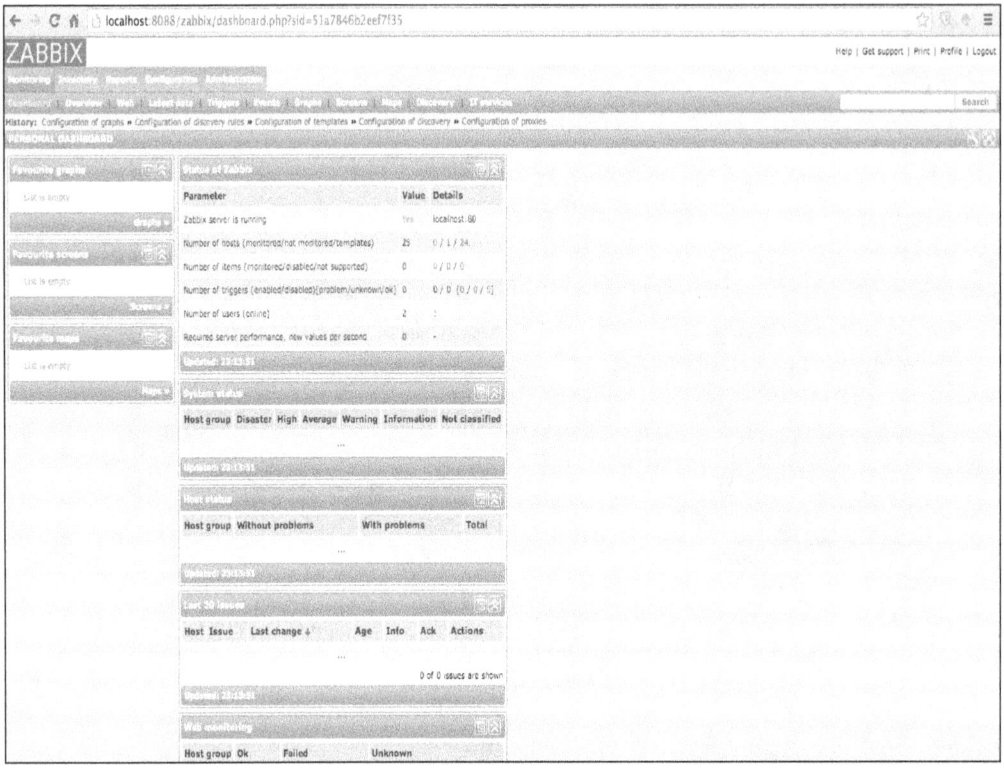

Zabbix web interface

Zabbix is also developed in an expandable way. So, it is easy to expand Zabbix with the provided plugins. As our topic is related to the monitoring of RabbitMQ, Zabbix also provides monitoring for the RabbitMQ with the help of plugins.

Before integrating RabbitMQ with Zabbix, we will download the source code of the plugin from its source Github using `git`. Then, we will copy all of the RabbitMQ-related scripts to the Zabbix external `script` folder, as shown in the following command line:

```
git clone https://github.com/adamlc/zabbix-rabbitmq.git
cd zabbix-rabbitmq
cp zabbix_* /ust/lib/zabix/externalscripts/
```

Now, we are ready to configure the plugin with our custom properties such as `hostname`, `username`, `password`, and so on as seen in the following source code:

```
// Zabbix Configuration
define('ZABBIX_HOSTNAME', 'localhost');

// RabbitMQ Configuration
define('API_HOSTNAME', 'localhost');
define('API_PORT', 15672);
define('API_USER', 'guest');
define('API_PASS', 'guest');
```

Finally, we need one configuration that is importing the plugin template to Zabbix to monitor RabbitMQ, as shown in the following screenshot:

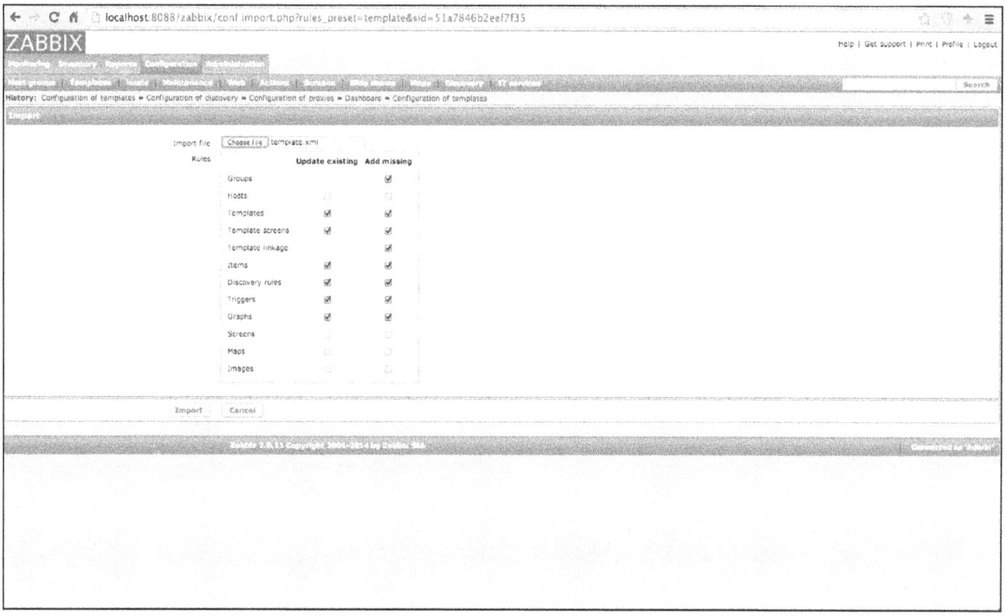

Importing template of RabbitMQ plugin of Zabbix

Summary

We have lots of software systems and integrations with other software systems in our technology stack now. Therefore, it is really important to control, manage, and monitor each of the software system in our server instances. Monitoring is crucial to check the errors and resource usages in our server instances. We have gone over software systems that provides monitoring capabilities.

RabbitMQ should be monitored since the main integration between systems can be provided with RabbitMQ. Monitoring RabbitMQ is quite easy with the provided tools called rabbitmqctl and the management plugin and also with the open source tools such as Nagios, Munin, and Zabbix. In the next chapter, we'll cover authentication and security.

8
Security in RabbitMQ

In worldwide computing areas, computer security or information security, which is known as cyber security has gained more importance than ever. Gartner reported that worldwide security expenses are increased by about 9% from 2012 to 2013.

As it is crucial to secure all software systems that we have, we should secure our message brokers too. As brokers have information about many parts of the dependent components, it's almost a must to secure RabbitMQ.

In this chapter, we will talk about the general vulnerabilities of the RabbitMQ servers and how we can solve these kinds of problems. After that, we will talk about the security mechanisms in the RabbitMQ, such as access control, SASL authentication, and SSL support as the following list shows:

- An introduction to security in RabbitMQ
- Access control
- SASL authentication
- SSL support in RabbitMQ

An brief introduction to security in RabbitMQ

Every server software system is allowed to access different types of software clients. Additionally, some software systems are allowed to access clients through network connections. Therefore, we should ensure the security of information behind the server software systems.

RabbitMQ has properties to configure security easily. Yet, every server application has some vulnerability. Therefore, we should use both RabbitMQ's solutions for the security issues and common solutions for server software systems.

Through this chapter, we'll dive deep into the vulnerabilities and their solutions for RabbitMQ server instances.

Vulnerabilities

The vulnerabilities of RabbitMQ server instances are similar to that of any standard server system. So, it is beneficial to list the current vulnerabilities of the server systems that are related to RabbitMQ. A complete report for vulnerabilities of Cenzic is published in Cenzic's Vulnerability Report 2014: `http://www.cenzic.com/downloads/Cenzic_Vulnerability_Report_2014.pdf`

Information leakage

Information leakage is, simply put, an application that inappropriately discloses sensitive data, such as messages of the message brokers, in our perspective, RabbitMQ. So, we have to ensure the security of the message details and its integrity.

Session management

Session management is simply defined as an application that inappropriately allows attackers to interject themselves as a logged in user of the software system. Therefore, we have to control our session management systems to block invalid users.

Authentication and authorization

As you probably already know, logging on to any computer with some credentials is authentication. Authorization is the process of verifying that you have access to something.

Vulnerability for authentication and authorization is simply defined as an application that does not properly ensure for unbreachable and unreplayable authentication and authorized access to data. Hence, authentication and authorization are to be properly enforced on the server side of the application. This includes enforcement of proper encrypted communication of credentials, password standards enforcement, feature and data access, ACL enforcements, and so on.

Message Brokers have authentication and authorization mechanism in their structure. Therefore, they have to ensure the security of the authentication and authorization of their own systems.

Solutions to the vulnerabilities

After going over related problems for the RabbitMQ, we are now ready to solve these problems using security technologies. As we know, without developing solutions for the vulnerabilities, our messages are not secure and are allowed to access from anyone who accesses our servers. Therefore, our main concern is to find solutions for the given security problems.

Fixing information leakage

The solution to Information Leakage is protecting documents from unauthorized people. As message brokers are highly data-oriented software systems, we have to be careful about the information leakage. The following list describes the general principles of preventing information leakage:

- Setting passwords to protect against unauthorized people
- Erasing or encrypting the information from leaking out
- Encrypting the messages within message brokers, where encryption is the process of encoding messages
- Limiting the usage of managing the software system

These properties have to be provided to prevent information leakage from the message broker.

Session management

Session Management is important for administering the message brokers and accessing message brokers as clients. We authenticate through the message broker's session management to send and receive messages. Therefore, we should ensure that the session management cannot be hijacked. The most powerful solution to prevent the hijacking of the session management is transmitting it over an encrypted protocol. One of the well-known encrypted protocols is **Secure Sockets Layer** (**SSL**). As a result, we have to use SSL to prevent session management hijacking.

Authentication and authorization

Authentication and authorization security is mostly related to the security of the session management. We should ensure the security of these systems using a secure protocol. As we discussed earlier in Session Management, we have to use SSL protocol to encrypt all the data communication.

Applying access control

RabbitMQ tries to solve problems, which we covered in the previous chapters, with the help of its mechanisms and plugins. **Access Control** simply specifies the permissions of the user within the virtual host. Each user has a different permission for each virtual host, for instance, a user a has *Read* permission on the virtual host TestVH.

We can manage access control using the `rabbitmqctl` command-line tool. The Rabbitmqctl tool gives us an opportunity to list all of the permissions of the user, as shown in the following command line:

```
vagrant@precise32:~$ sudo rabbitmqctl list_user_permissions guest
Listing permissions for user "guest" ...
/.*    .*.*
...done.
```

Moreover, we have another chance to list permissions of the provided virtual host, as shown in the following command line:

```
vagrant@precise32:~$ sudo rabbitmqctl list_permissions -p /
Listing permissions in vhost "/" ...
guest.*.*.*
monit.*.*.*
monitor.*.*.*
...done.
```

Furthermore, we can delete all of the permissions of the user within a provided virtual host. The following command line shows the clearing permissions of the user `jack` within the root virtual host:

```
vagrant@precise32:~$ sudo rabbitmqctl clear_permissions -p / jack
Clearing permissions for user "jack" in vhost "/" ...
...done.
```

Finally, we need to set new permissions for the user within the provided virtual host. As you can see in the following command line, we can set new permissions to the user jack within the root virtual host:

```
vagrant@precise32:~$ sudo rabbitmqctl set_permissions -p / jack ".*" ".*"
".*"
Setting permissions for user "jack" in vhost "/" ...
...done.
```

Rabbitmqctl is a great tool for managing the RabbitMQ server and it is enough for us to manage access control. Additionally, some administrators would like to use graphical interfaces for administrating the RabbitMQ server. We have another use of RabbitMQ's management plugin to list each user's permission and change their permissions for the provided **Virtual Host**, as shown in the following image:

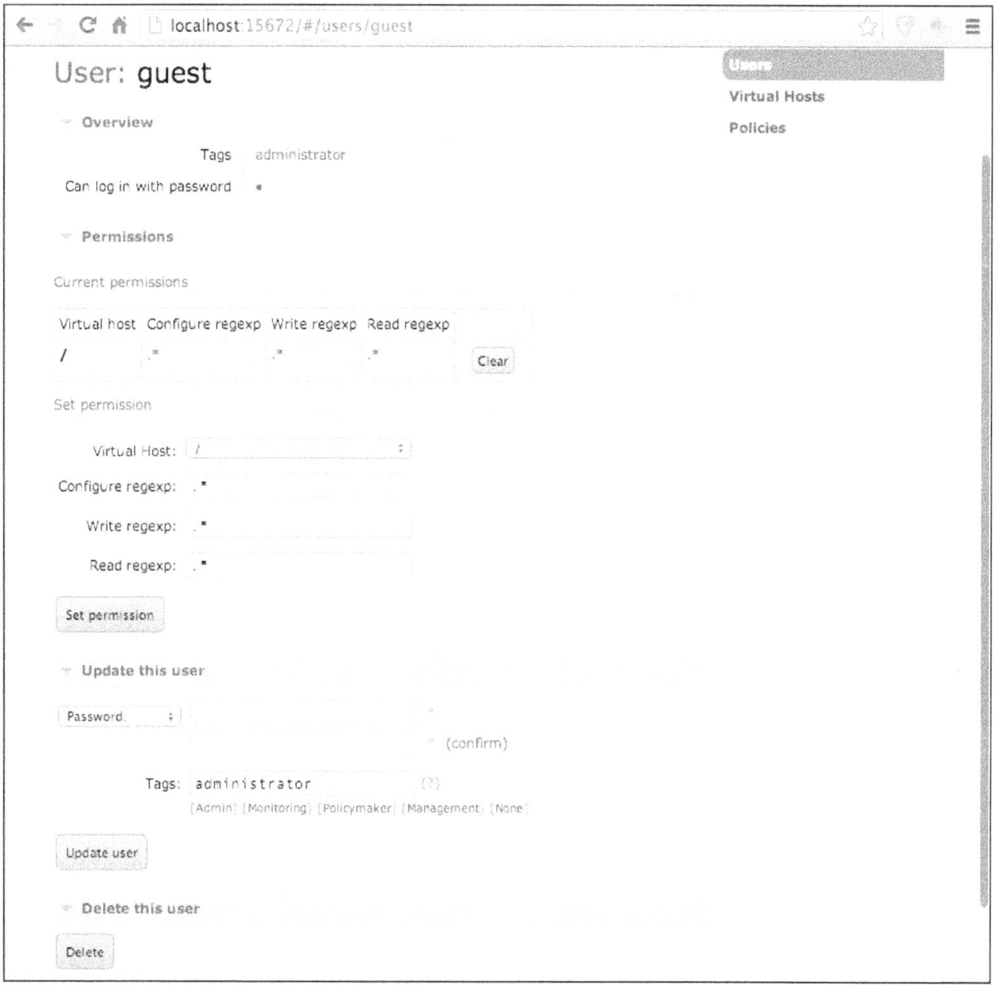

Figure:8.1: Access control of user

As a result, we should limit the user's activities within the provided virtual host. Next, we move on to access control, which manages the user permissions for the provided virtual host.

Providing SASL authentication

The **Simple Authentication and Security Layer (SASL)** is a framework for providing authentication and data security services in connection-oriented protocols via replaceable mechanisms according to the official SASL protocol specification. SASL specifies the structured interface between protocols and mechanisms. As SASL is a framework on top of the other frameworks, we can use SASL into SMTP, LDAP, XMPP, and other communication protocols. SASL provides the abstraction layer for each of the communication protocols, as shown in the following image:

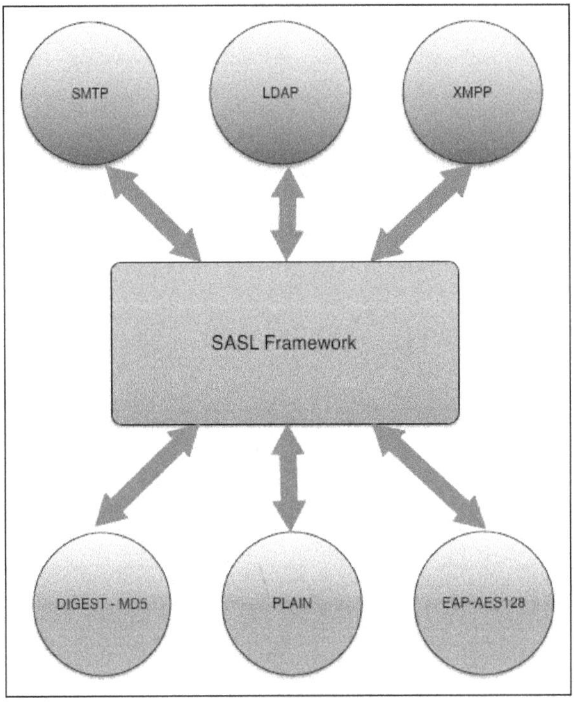

Figure 8.2: SASL framework

RabbitMQ has a plugin to support SASL authentication mechanisms. There are three mechanisms built into the server:

- **Plain**: SASL PLAIN authentication provided. This is enabled by default in the RabbitMQ server and clients.
- **AMQPlain**: This is nonstandard version of PLAIN that is defined in the AMQP 0-8 specifications.
- **Rabbit-CR-Demo**: This is the nonstandard mechanism, which demonstrates challenge-response authentication, is provided according to the RabbitMQ documentation.

- **External**: Custom mechanism to externally control authentication.
- **SSL**: In this case, the external mechanism is SSL, which can authenticate users with certificate
- **LDAP:** In this case, the external mechanism is LDAP, which can authenticate users using the LDAP database
- **HTTP:** In this case, the external mechanism is HTTP, which can authenticate users from web server that knows the user credentials.

Before using the SASL plugin in your RabbitMQ server, you should choose the mechanism in the configuration file. The default value of the `auth_mechanisms` key is `['PLAIN', 'AMQPLAIN']`; however, you can change the default values according to your credential system.

Additionally, if you define an authentication mechanism for SASL, you should connect to the servers from the client's setting `SaslConfig` properties of the API. In Java API, you can find the current SASL config with the function `getSaslConfig` of `ConnectionFactory`. Moreover, you can set the related mechanism using the `ConnectionFactory.AuthMechanisms` object within C# API.

SSL support in RabbitMQ

Secure Sockets Layer (SSL) is a standard protocol for establishing an encrypted link between a web server and a browser. SSL uses a cryptographic system that uses two keys to encrypt data; one is public key known to everyone and the other is a private key known only to the recipient of the message.

Keys, certificates, and CA certificates

OpenSSL is a software library to be used in applications that need to secure communications. It has been widely adopted by the users for Internet web servers. The library consists of open source implementations of SSL and TSL as well as basic cryptographic functions. With the help of OpenSSL, RabbitMQ can establish an encrypted communication channel and exchange signed certificates. In order to verify a certificate, a chain of trust for certificates should be formed. The last element of the chain is the root certificate, which is a self-signed certificate.

Enabling SSL support

If we look at the SSL support in RabbitMQ, it has a built-in support for SSL; however, we need to enable SSL support in RabbitMQ. To enable the SSL/TLS support in RabbitMQ, we should provide the location of root certificate, the server's certificate file and the server's key. We can provide the locations of these files in the configuration file as shown in the following code snippet:

```
[
{rabbit, [
{ssl_listeners, [5671]},
{ssl_options, [{cacertfile, "/path/to/pathca/cacert.pem"},
{certfile, "/path/to/serverpath/cert.pem"},
{keyfile, "/path/to/serverpath/key.pem"},
{verify, verify_peer},
{fail_if_no_peer_cert, false}]}
]}
]
```

When you look at the details of the file snippet, you can find the location of the certification files (`/path/to/serverpath/cert.pem`) and key files(`/path/to/serverpath/key.pem`). Moreover, you can change the port of the SSL listener using the `ssl_listeners` attribute.

In client programming, you should load the key and certification file to establish in a trusted connection between RabbitMQ servers and clients. This issue will be discussed in detail in the next chapters.

As a consequence, SSL is the de facto standard for securing the HTTP protocol. So, it is really secure to use SSL support in RabbitMQ to ensure the security of all communications between the clients and server.

Summary

Nowadays, we have lots of hacker attacks, such as distributed attacks, on our software systems within newer technologies. So, we should ensure that we know all our vulnerabilities to make our software system secure.

RabbitMQ is a server software, so it communicates with many clients. Therefore, most of the server security problems are seen in the RabbitMQ. In order to ensure the security in RabbitMQ, we had different type of authentications and securing the protocol itself. With the help of SASL and SSL support, we provided security of our messages and communication.

In the next chapter, we will talk about developing clients with real-world examples in different programming languages.

9

Java RabbitMQ Client Programming

We have discussed the internal structure of the RabbitMQ server, and managing, monitoring, and so on of RabbitMQ server instances. Now, we are ready to dive into the details of the development of clients using RabbitMQ in different languages starting with Java.

RabbitMQ has clients for Java, Python, C#, Ruby, and so on. Using these clients, messaging applications can be implemented in the simplest way. As we have talked about the details of AMQP and how it's implemented in RabbitMQ in *Chapter 3, Architecture and Messaging*, we can now implement all of the messaging capabilities such as direct messaging, pub-sub messaging, routed messaging, and other messaging capabilities.

To understand these capabilities thoroughly, we chose to develop our clients according to a case study. Our case study is called **Collaborative Application**, where messaging can be used extensively. We'll talk about the details of the case study in this chapter. Moreover, we'll talk about the basics and the details of the RabbitMQ Java Client API and developing our clients according to the case study's use cases. Later on, we will discuss the Spring framework integration and Spring AMQP. The following list shows the main themes of the present chapter:

- Case study
- Java
- Java Message Service (JMS)
- RabbitMQ Java client API
- Case study client implementations
- Spring framework and RabbitMQ
- Spring AMQP

Case study

The chapters from here on have the goal of teaching you how to develop clients for RabbitMQ. The easiest way to learn developing software is using a case study. Therefore, we chose a case study that is called **Collaborative Software.**

According to Wikipedia, collaborative software is an application designed to help people involved in a common task to achieve goals. In collaborative software, people engage each other in real-time. They share documents, images, and other types of files. Moreover, they talk to each other. Sometimes, managers would like to talk with a specific person; and some other times, they would prefer talking with a group of people. As a result, our collaboration software should integrate many software systems and it is heavily based on messaging. Then, it comes to RabbitMQ that integrates and scales the messaging facilities.

Before developing our collaborative software, we need to clarify our requirements for our software using a use case diagram and interaction diagram (sequence diagram). Then we are ready to develop our collaboration software according to our software requirements and designs.

Use cases

Before developing the case study called collaboration software, we should look at the requirements of our app. And for this, it is good to think about and design use cases. As we analyze other collaboration applications, we will see that all apps have some common features. These features might be sending the message to a single user, sending group messages, sending bulk messages, sending file messages, or creating some tasks such as changing the image format or parsing the documents to search. We can find these use cases in the following use cases screenshot:

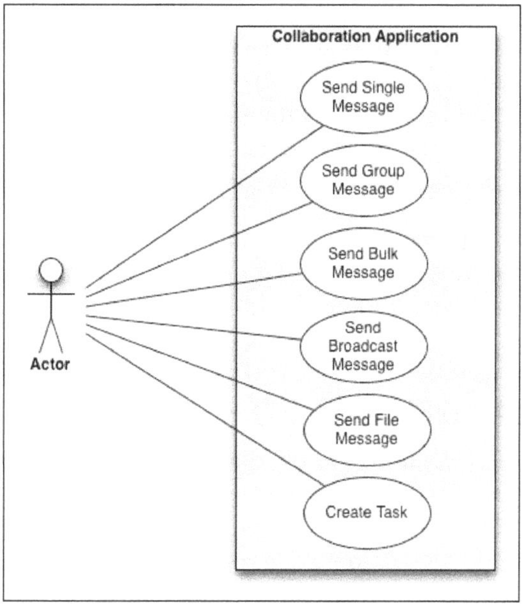

Use Cases Diagram

Interaction diagram – sequence diagram

After defining the use cases, we should describe the sequence of actions throughout each use cases. Then it come to interaction diagrams, which can picture a control flow with nodes. We can describe the sequence of each use case with a sequence diagram, which is an interaction diagram that shows how processes operate with one another and in what order. For example, the following screenshot of a sequence diagram clearly shows how a single message is sent to the right user:

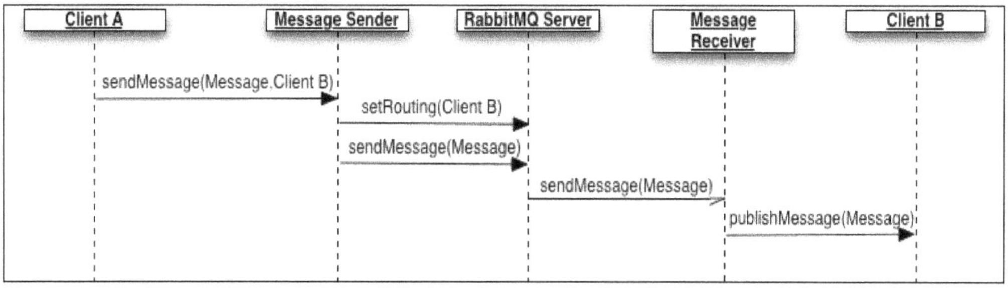

Direct Message Sequence Diagram

In the preceding screenshot, we can see that our **Client A** sends a message to **Client B** using **Message Sender**. After that, a **Message Sender** object sets the routing key of the **Message Broker**, which is RabbitMQ in our case, then sends the message through the **RabbitMQ Server**. RabbitMQ finds the right client using the routing key, then sends the message to the message listener called **Message Receiver**. Finally, **Message Receiver** publishes the message to the attached client, which is **Client B** in our case.

As a consequence, our sequence within each use case behaves like the previous sequence diagram. From now on, we will focus on explaining the technologies that we will use in the current chapter and the implementation of our case study.

Application language – Java

Java is a well-known, widely accepted by enterprises, and currently one of the most popular languages of our modern software systems. Moreover, Java is not only a programming language; it also gives a great number of libraries that form enterprises, mobiles, and web platforms. Therefore, we call Java a platform.

Java was developed by the company **Sun** that has now been acquired by **Oracle**. Now Java's specifications are determined from Oracle and open source community. As Java is the first choice for enterprise platforms, we aim to talk about RabbitMQ in Java first. To be more specific, the Java platform has a messaging specification called JMS within the Enterprise Edition of Java Platform. We will talk about the details of the JMS and its similarities and differences between RabbitMQ, Java clients, and JMS in the following topic. The following is the Java's logo:

Java Platform

Java Message Service (JMS)

As our main concern is messaging and message brokers, we firstly look at the Java Platform's answer to well-known messaging problems. **Java for Enterprise Edition (J2EE)** framework supplies messaging protocol for messaging applications called **Java Message Service (JMS)**.

The JMS is a Java API that allows applications to create, send, receive, and read messages. JMS defines the common set of interfaces that allow Java applications to communicate with each other. We can list the properties of JMS as follows:

- Asynchronous
- Reliable
- Loosely coupled

The general structure of JMS applications can be seen in the following screenshot:

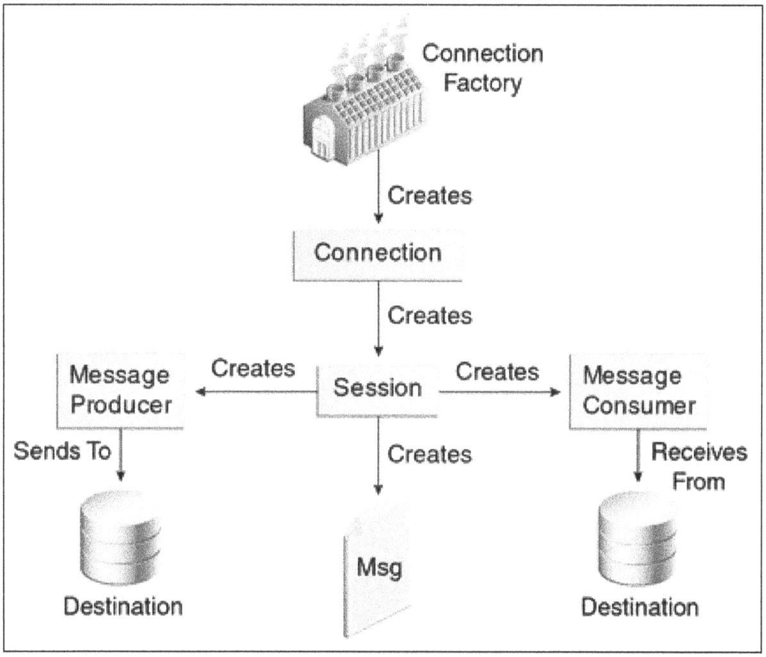

Java Message Service

JMS has connections that create sessions. Sessions are able to handle the production of messages and consumption of messages by using defined interfaces within the JMS API.

JMS is commonly used to message API for Java Platforms. Therefore, we can use the messaging facilities of the JMS within the Java applications. However, nowadays we develop applications using different languages and different platforms, so, we should integrate different types of software systems. This brings us to AMQP, that allows integrating different types of software systems.

On the other hand, many of the modern Java applications use JMS as a messaging structure. However, we need a more feature rich and portable distributed messaging software than JMS. RabbitMQ and its protocol AMQP give you more advantages than using the JMS with respect to message routing skills, message model skills, and so on. Therefore, it is also good to use RabbitMQ as a message broker, and it could be used within the JMS clients.

Lastly, RabbitMQ has a JMS API that was developed by **VMware**, and it could be used with a commercial license. This API is well supported and it can be used with RabbitMQ on top of JMS.

RabbitMQ Java client API

The RabbitMQ community and its main supporter company, **Pivotal**, provide an official client library for Java called **RabbitMQ Java Client**. Client library provides both the publishing of messages and receiving of messages. Moreover, Client library supports both synchronous receiving and asynchronous receiving. The details will be explained in the following topics.

If we look at the main packages of the RabbitMQ Java Client, we can see three packages as shown in the following screenshot:

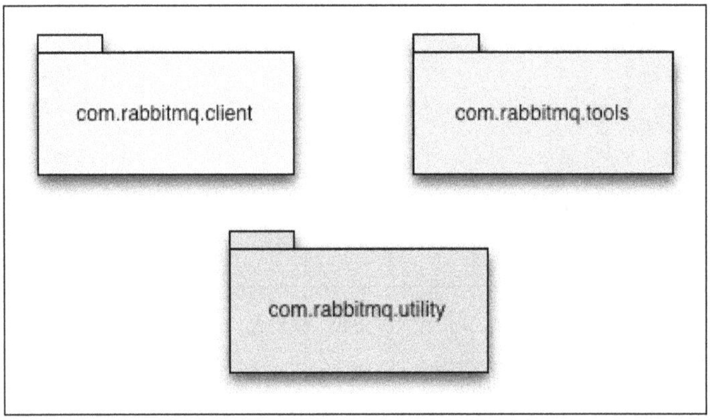

RabbitMQ Java Client Packages

Let's take a look at the following explanation:

- **com.rabbitmq.client** package provides classes and interfaces for AMQP connections, channels, and wire-protocol framing descriptions
- **com.rabbitmq.tools** provides classes and methods for non-core utilities and administration tools
- Lastly, **com.rabbitmq.utility** provides helper classes which are mostly used in the implementation of a library

As the most important package of the API is Client API, we cover the basics and the internals of the API in the following topics:

Client package in detail

You can find each AMQP element in the Client package such as Connection, Channel, Exchanges, Queues, and so on. You can find each functionality of AMQP in the Client package too. Therefore, we would like to introduce you to each element and its functions as follows:

Connection

Connection is an interface in the Client package. `Connection` interface directly refers to the `Connection` element of AMQP. So, `Connection` interface covers the functionalities of the `Connection` element of AMQP too. We can create a `Connection` instance through the `ConnectionFactory` class as shown in the following code:

```
//ConnectionFactory initialization
ConnectionFactory factory = new ConnectionFactory();
//Setting the hostname
factory.setHost("localhost");
//Setting the Username
factory.setUsername("guest");
//Setting the Password
factory.setPassword("guest");
//Creating the connection using factory instance
Connection conn = factory.newConnection();
```

The `ConnectionFactory` class has attributes that refer to `hostname`, `port`, `username`, `password`, and `virtual host`. We can set each of the mandatory attributes, and then we are ready to create our connection. Additionally, we can set each attribute using the URI standards as follows:

```
//ConnectionFactory initialization
ConnectionFactory factory = new ConnectionFactory();
//Setting the Attributes using Uri
factory.setUri("amqp://guest:guest@localhost")
//Creating the connection using factory instance
Connection conn = factory.newConnection();
```

As we recall from the database connections, we should close our connection after completing our tasks. In `Connection` class, we have the `close()` method to do the same job as the following code example demonstrates:

```
//Creating the connection using factory instance
Connection conn = factory.newConnection();
//Closing the connection
conn.close();
```

Channel

Channel is another interface in the Client package. As Connection interface refers to the `Connection` element of AMQP, Channel refers to the `Channel` element of AMQP. As we discussed in *Chapter 3, Architecture and Messaging*, Channel's main role is to serve as a logical connection inside of the network connection to the message broker. `Channel` instances are thread safe.

The `Channel` instance could be initialized through the `Connection` instance as shown in the following code example:

```
//Connection Init
Connection conn = factory.newConection();
//Initializing the Channel using Connection
Channel channel = conn.createChannel();
```

Because of the Channel's main responsibility, we can send message, receive message, make queue operations, and so on using the Channel. Channel won't be available if these operations fail. Code examples of channel could be presented in the following topics.

Exchanges

Exchanges are the main elements of AMQP that moderate the queues with given functionalities. Exchanges are also available within the RabbitMQ Java Client API. Exchanges' main responsibility is to receive the messages from producers and push them to the related queues that are expressed by the rules. Although they have such importance at AMQP 0.9.1, they don't exist in the AMQP 1.0 specification.

We are able to define each exchange type such as direct, fanout, headers, and so on using the Java API. In Java API, we can create the exchanges via `Channel` instances as shown in the following code example:

```
//Channel Initialization
Channel channel = conn.createChannel();
//Declaring Exchanges using the Channel
channel.exchangeDeclare("mastering.rabbitmq","fanout");
```

Queues

Message Brokers are nothing without queues. Queues are the most important part of the Message Brokers and AMQP. Whenever a new message consumer or subscriber is connected to the Exchange, RabbitMQ creates a queue for the related exchange with the provided name.

As we discussed earlier, Channels are responsible for common operations of the Queues. Therefore, we can declare, bind, unbind, purge, and delete queues with the methods of the Channels as provided in the RabbitMQ Java API. The following simple coding example shows how Queues are bound to given exchanges:

```
//Declare Exchange
channel.exchangeDeclare("mastering.rabbitmq", "fanout")
//Get the name of bound Queue
String queueName = channel.queueDeclare().getQueue()
//Bind the queue to the exchange without routing key
channel.queueBind(queueName, "mastering.rabbitmq","");
```

Publishing messages

Before talking about the details of sending messages within our case study, we should look at how we send messages through RabbitMQ using RabbitMQ Java Client API. Although we know that we have many methods such as pub-sub, routed messaging, and so on to publish our message, we'd just like to show the simple message sending.

As we'd like to show simple message sending, we should declare queue and publish message to the declared queue as shown in the following code example:

```
importcom.rabbitmq.client.ConnectionFactory;
importcom.rabbitmq.client.Connection;
importcom.rabbitmq.client.Channel;

publicclassSender {

  privatefinalstatic String QUEUE_NAME ="mastering.rabbitmq";

  publicstaticvoidmain(String[] argv)throws Exception {

    ConnectionFactory factory =new ConnectionFactory();
    factory.setHost("localhost"); //1
    Connection connection = factory.newConnection(); //2
    Channel channel = connection.createChannel(); //3

    channel.queueDeclare(QUEUE_NAME,false,false,false,null);//4
    String message ="Hello Mastering RabbitMQ!";
    channel.basicPublish
    ("",QUEUE_NAME,null,message.getBytes());//5
    System.out.println("Following Message Sent: "+ message);

    channel.close();
    connection.close();
  }
}
```

If we look at the details of the code that has numbered comments, we notice that:

- Connection Factory expresses the hostname of RabbitMQ Server
- The `Connection` instance created through `ConnectionFactory` instance
- The `Channel` instance initialized through the `Connection` instance
- Declaring a queue with a provided name
- Publishing message directly to the provided queue

Now, we've got the basics of the sending message, we're ready to move on to how we receive a message that delivers from the connected queue.

Consuming messages

Consuming messages from RabbitMQ is similar, but not identical to publishing. Firstly, we initialize the connection through the `ConnectionFactory` instance and declare the queue that is related to our receiver and sender. Then, the difference from sender comes here, that is, the receiving message part. The receiving part could be implemented in a synchronous way or asynchronous way. In a synchronous way, we block the current thread to listen to message deliveries; however, in an asynchronous way, a thread can't be blocked, so whenever a message is delivered, the consumer method is called instantly in an event like manner.

Synchronously receiving messages

The consumer can receive messages synchronously, and we will go over an example regarding this. As we look at the following code example, we block our thread to listen to message deliveries by using the `while` loop. In a `while` loop, we fetch the next incoming message using the `QueueinConsumer` instance. Then we can convert the incoming message body to our custom object type:

```
importcom.rabbitmq.client.ConnectionFactory;
importcom.rabbitmq.client.Connection;
importcom.rabbitmq.client.Channel;
importcom.rabbitmq.client.QueueingConsumer;

publicclassReciever {

  privatefinalstaticStringQUEUE_NAME="mastering.rabbitmq";

  publicstaticvoidmain(String[]argv)throwsException {

    ConnectionFactoryfactory=newConnectionFactory();
    factory.setHost("localhost");
    Connectionconnection=factory.newConnection();
    Channelchannel=connection.createChannel();

    channel.queueDeclare(QUEUE_NAME,false,false,false,null);

    QueueingConsumerconsumer=newQueueingConsumer(channel);
    channel.basicConsume(QUEUE_NAME,true,consumer);

    while(true) {
      QueueingConsumer.Deliverydelivery=consumer.nextDelivery();
      Stringmsg=newString(delivery.getBody());
      System.out.println("Received Message:"+msg);
    }
  }
}
```

Asynchronously receiving messages

The main functional difference between synchronous receiving and asynchronous receiving is blocking. In asynchronous receiving, thread couldn't be blocked by the listening part, so you can do anything with the current thread.

Non-blocking and event driven software systems are very popular for their scalability. Therefore, it is good to use an asynchronous way in the receiving part. RabbitMQ Java API gives us a `DefaultConsumer` method to control the deliveries. In the following code example, you can find the inner class that implements the `DefaultConsumer` method called `handleDelivery`:

```
importcom.rabbitmq.client.Connection;
importcom.rabbitmq.client.Channel;
importcom.rabbitmq.client.QueueingConsumer;

publicclassReciever {

  privatefinalstaticStringQUEUE_NAME="mastering.rabbitmq";

  publicstaticvoidmain(String[]argv)throwsException {
    ConnectionFactoryfactory=newConnectionFactory();
    factory.setHost("localhost");
    Connectionconnection=factory.newConnection();
    Channelchannel=connection.createChannel();

    channel.queueDeclare(QUEUE_NAME,false,false,false,null);

    channel.basicConsume(QUEUE_NAME, false, new
    DefaultConsumer(channel) {
      @Override
      public void handleDelivery(String consumerTag, Envelope
      envelope,
      AMQP.BasicProperties properties, byte[] body)throws
      IOException {
        String msg = new String(body);
        System.out.println("Received Message: " + msg);
      }
    });
  }
}
```

Case study – client implementations

After showing the basics of our RabbitMQ Java Client API, we are now ready to implement our collaboration application. In our collaboration application, we have different functions that are well-defined in the use cases.

We'd like to focus on the messaging part, instead of focusing on all the parts, to learn how to send and receive messages between our systems. Before diving into the messaging parts, we'd like to introduce you to **model classes** that define our **Message** class instances.

Model classes

In Model-View-Controller architecture, models are responsible for containing the business logic. With the same idea, our model classes contain message logics such as Message class which defines a simple message.

In messaging systems, you have to communicate through the binary format. Therefore, we should use one of the serializing and de-serializing mechanisms. In our collaboration application, we are going to use **JavaScript Object Notation (JSON)** for its great support for easily serializing and de-serializing. There are many JSON serialization libraries like **Kryo** and **ProtoBuf**.

JSONMessage interface

As we picked the JSON format to communicate within RabbitMQ sender and receivers, we should append JSON format to `String` method in each message `model` classe. We can ensure that each messaging model has JSON to String method by implementing the `JSONMessage` interface that is defined in the following code example:

```
package com.collaboration.model.json;

publicinterface JSONMessage {
  /**
   * @return String
   */
  public String toJSON();
}
```

Message model

Message model simply expresses who sends the message, called the "from" variable in our class; who receives the message, called the "to" variable in our class; header; content; and message number that denotes the unique message number sequence.

Moreover, we need two methods: serializing to JSON String from JSON Object, and de-serializing to Java objects from JSON String. We call them `toJSON` and `fromJSON` methods. We make use of Google's JSON library for serialization.

Message model class is our main class that we use mostly in our collaboration application. The following code shows the Message model class:

```
package com.collaboration.model;

import com.collaboration.model.json.JSONMessage;
import com.google.gson.Gson;

/**
 * @author Emrah Ayanoglu
 * Following code represents the simple message model
 *
 */
public class Message implements JSONMessage {
  private int msgNo;
  private String from;
  private String to;
  private String header;
  private String content;

  private static Gson gson = new Gson();

  public String toJSON() {
    return gson.toJson(this);
  }

  public static Message fromJSON(String msg) {
    Gson gson = new Gson();
    return (Message) gson.fromJson(msg, Message.class);
  }
```

```
  public String toString() {
    return String.format("Message No: %d From: %s To: %s " +
    "Header: %s Content: %s",
    msgNo, from, to, header, content);
  }
}
```

File message

Although most of our messages have text contents, we also need file messages to send our documents, pictures, and other important files to our group members in our collaboration application. We should change the type of our content variable from String to byte array to store the file content. The other variables remain the same:

```
package com.collaboration.model;

import java.io.UnsupportedEncodingException;

import com.collaboration.model.json.JSONMessage;
import com.google.gson.Gson;

/**
 * @author Emrah Ayanoglu
 * Following Code just represents the model of the message
 * that handles the binary messages
 */
public class FileMessage implements JSONMessage {
  private int msgNo;
  private String from;
  private String to;
  private String header;
  private byte[] file_content;

  public String toJSON() {
    Gson gson = new Gson();
    return gson.toJson(this);
  }

  public static FileMessage fromJSON(String msg) {
    Gson gson = new Gson();
    return (FileMessage) gson.fromJson(msg, FileMessage.class);
  }
```

```
    public String toString() {
      try {
        return String.format("Message No: % From: %s To: %s " +
        "Header: %s File Content: \n %s UTF-8",
        msgNo,from, to, header, getFile_content());
      }
      catch (UnsupportedEncodingException e) {
        e.printStackTrace();
        return "";
      }
    }
  }
```

Task

Lastly, we mostly use background tasks that take lots of time to process our documents, image processing, sending lots of e-mails at once, and so on. RabbitMQ gives us another opportunity to control the background tasks too. Therefore, we need to provide another message model class that expresses the tasks.

A Task class instance should have an ID that defines the unique identifier, hostname of the application, delivery info, command that defines the command to be executed, errback that expresses the error information, and lastly, expires that simply defines the expiration time of the task as seen in the following example:

```
package com.collaboration.model;

import java.util.Date;
import java.util.UUID;

import com.collaboration.model.json.JSONMessage;
import com.google.gson.Gson;

/**
 * @author Emrah Ayanoglu
 *
 * Following code represents the task message model that
 * encapsulates the related information of task
 */
public class Task implements JSONMessage {
  private String id;
  private String hostname;
  private String delivery_info;
  private String command;
  private String errback;
```

```
    private Date expires;

    public Task() {
      id = UUID.randomUUID().toString();
      hostname = "";
      delivery_info = "";
      command = "";
      errback = "";
      expires = new Date();
    }

    public static Task fromJSON(String msg) {
      Gson gson = new Gson();
      return (Task) gson.fromJson(msg, Task.class);
    }

    public String toJSON() {
      Gson gson = new Gson();
      return gson.toJson(this);
    }

    public String toString() {
      return "ID: " + getId() + " Hostname: " + getHostname()
      + " Delivery Info: " + getDelivery_info() + " Callbacks: "
      + getCommand() + " Expires: " + getExpires().toString();
    }
  }
```

Single message

Our first use case of the collaboration app is to send the single message from one user to another specific user. The common example of single message is private messaging between two users. To accomplish this kind of use case, we just need to send our message directly to the queue, which is bound to another user. Therefore, our sender just connects to the right queue and enqueues the message, and our receiver listens to the queue and dequeues the message from the queue, as seen in the following screenshot:

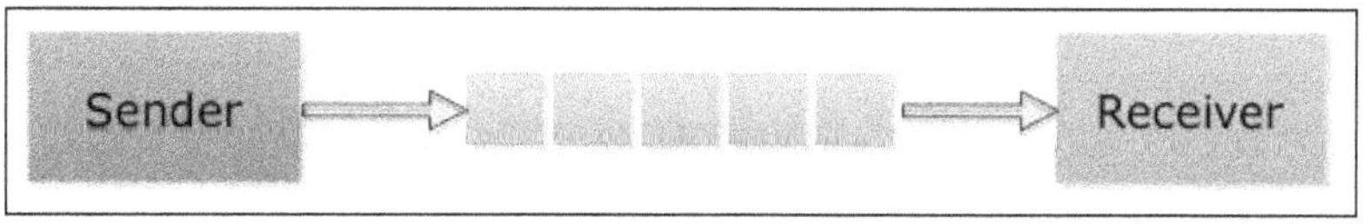

Single Message Architecture

Sender

As said earlier, our sender creates the connection to the RabbitMQ Server and declares the queue that is bound to another client. Then, sender creates the `Message` objects and serializes into the JSON format. Finally, our sender converts the JSON string to the binary array and sends it to the queue, as shown in the following code:

```
package com.collaboration.sender;

import java.io.IOException;

import com.collaboration.model.Message;
import com.rabbitmq.client.Channel;
import com.rabbitmq.client.Connection;
import com.rabbitmq.client.ConnectionFactory;

publicclass Sender {
  privatefinalstatic String QUEUE_NAME = "mastering_rabbitmq";

  /**
   * @param argv
   * @throws IOException
   */
  publicstaticvoid main(String[] argv) throws IOException {
    ConnectionFactory factory = new ConnectionFactory();
    factory.setHost("localhost");
    Connection connection = factory.newConnection();

    Channel channel = connection.createChannel();
    channel.queueDeclare(QUEUE_NAME, false, false, false, null);

    Message msg = new Message();
    msg.setFrom("John");
    msg.setTo("Nicky");
    msg.setHeader("Hello World");
    msg.setContent("Hello World Again");

    for(int i = 0; i < 5; i++) {
      msg.setMsgNo(i + 1);
      channel.basicPublish("", QUEUE_NAME, null,
      msg.toJSON().getBytes());
    }
```

```
        System.out.println("Message is sent: " + msg.toString());

        channel.close();
        connection.close();
    }
}
```

Receiver

Our receiver creates the connection to the RabbitMQ Server and declares the queue called `mastering_rabbitmq`. After this, we need to adopt our queue to listen to the incoming messages. To handle the incoming messages, we should listen to the incoming messages for all the time, which is handled in the blocking manner. Finally, we fetch the delivered message and de-serialize it from the string to `Message` object using JSON conversion:

```
package com.collaboration.receiver;

import java.io.IOException;

import com.collaboration.model.Message;
import com.rabbitmq.client.Channel;
import com.rabbitmq.client.Connection;
import com.rabbitmq.client.ConnectionFactory;
import com.rabbitmq.client.ConsumerCancelledException;
import com.rabbitmq.client.QueueingConsumer;
import com.rabbitmq.client.ShutdownSignalException;

public class Receiver {
  private final static String QUEUE_NAME = "mastering_rabbitmq";

  /**
   * @param argv
   * @throws IOException
   * @throws ShutdownSignalException
   * @throws ConsumerCancelledException
   * @throws InterruptedException
   */
  public static void main(String[] argv) throws IOException,
  ShutdownSignalException, ConsumerCancelledException,
  InterruptedException {
    ConnectionFactory factory = new ConnectionFactory();
    factory.setHost("localhost");
```

```
        Connection connection = factory.newConnection();
        Channel channel = connection.createChannel();

        channel.queueDeclare(QUEUE_NAME, false, false, false, null);

        System.out.println("Waiting for the messages.........");

        QueueingConsumer consumer = new QueueingConsumer(channel);
        channel.basicConsume(QUEUE_NAME, true, consumer);

        while (true) {
          QueueingConsumer.Delivery delivery =
          consumer.nextDelivery();
          String msg = new String(delivery.getBody());

          System.out.println("Received: " +
          Message.fromJSON(msg).toString());
        }
      }
    }
```

Group message – routing

Another important use case of our collaboration app is to send message to the specific group; for instance, we are a group of fans of a football team and our group manager would like to send a message to the related group. In RabbitMQ terms, we should use routed exchanges to send a group message. In routed exchanges, sender sends message to the specific exchange providing a topic, such as a group name. Then, all receivers within the same group are able to fetch the messages, as shown in the following screenshot:

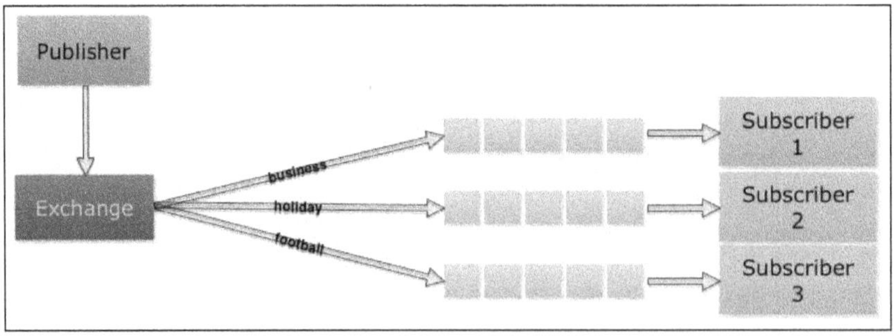

Group Message Architecture

Sender

Sender just sends messages to their group in a routed messaging. To make it possible, sender first connects to the RabbitMQ Server, and then declares exchange with the **topic** functionality. Finally, sender sends its message to the exchange with topic, which is "`*.business.*`" in our example. RabbitMQ exchanges fetch the message from the sender and enqueue message to each queue that is bound with the receivers. You can find the details of the sender in the following code example:

```
package com.collaboration.sender;

import java.io.IOException;

import com.collaboration.model.Message;
import com.rabbitmq.client.Channel;
import com.rabbitmq.client.Connection;
import com.rabbitmq.client.ConnectionFactory;

public class GroupSender {

  private final static String EXCHANGE_NAME =
  "mastering_rabbitmq_group";

  /**
   * @param args
   * @throws IOException
   */
  public static void main(String[] args) throws IOException {
    ConnectionFactory factory = new ConnectionFactory();
    factory.setHost("localhost");
    Connection connection = factory.newConnection();

    Channel channel = connection.createChannel();
    channel.exchangeDeclare(EXCHANGE_NAME, "topic");

    Message msg = new Message();
    msg.setFrom("John");
    msg.setTo("Nicky");
    msg.setHeader("Hello World");
    msg.setContent("Hello World Again");
```

```
        for(int i = 0; i < 5; i++) {
          msg.setMsgNo(i + 1);
          channel.basicPublish(EXCHANGE_NAME, "*.business.*", null,
          msg.toJSON().getBytes());
        }

        System.out.println("Message is sent: " + msg.toString());

        channel.close();
        connection.close();
      }
    }
```

Receiver

Receiver listens to the incoming messages in the queue. Before, while listening to the queue, we do the exact same thing with the sender. We connect to the RabbitMQ Server and declare our exchanges with topic functionality, and bind to the queue with the provided topic that is related with the group name. Then we are ready to listen to the upcoming messages. You can follow the receiving implementation with the following code example:

```
package com.collaboration.receiver;

import java.io.IOException;

import com.collaboration.model.Message;
import com.rabbitmq.client.Channel;
import com.rabbitmq.client.Connection;
import com.rabbitmq.client.ConnectionFactory;
import com.rabbitmq.client.ConsumerCancelledException;
import com.rabbitmq.client.QueueingConsumer;
import com.rabbitmq.client.ShutdownSignalException;

public class GroupReceiver {
  private final static String EXCHANGE_NAME =
  "mastering_rabbitmq_group";

  /**
   * @param args
```

```
     * @throws IOException
     * @throws ShutdownSignalException
     * @throws ConsumerCancelledException
     * @throws InterruptedException
     */
    public static void main(String[] args) throws IOException,
    ShutdownSignalException, ConsumerCancelledException,
    InterruptedException {
      ConnectionFactory factory = new ConnectionFactory();
      factory.setHost("localhost");

      Connection connection = factory.newConnection();
      Channel channel = connection.createChannel();

      channel.exchangeDeclare(EXCHANGE_NAME, "topic");
      String queueName = channel.queueDeclare().getQueue();
      channel.queueBind(queueName, EXCHANGE_NAME, "*.business.*");

      System.out.println("Waiting for the messages.........");

      QueueingConsumer consumer = new QueueingConsumer(channel);
      channel.basicConsume(queueName, true, consumer);

      while (true) {
        QueueingConsumer.Delivery delivery =
        consumer.nextDelivery();
        String msg = new String(delivery.getBody());

        System.out.println("Received: " +
        Message.fromJSON(msg).toString());
      }
    }
  }
```

Bulk message – PubSub

Another important use case is sending a bulk message. The difference between bulk message and routed message is that, bulk message sends a message to all of the clients; however, routed message sends messages to the group of clients that are defined.

In message broker terms, bulk message is defined as **PubSub**, which is the abbreviation of publish and subscribe. Sender behaves like a publisher, which publishes messages to all the receivers that are called subscribers:

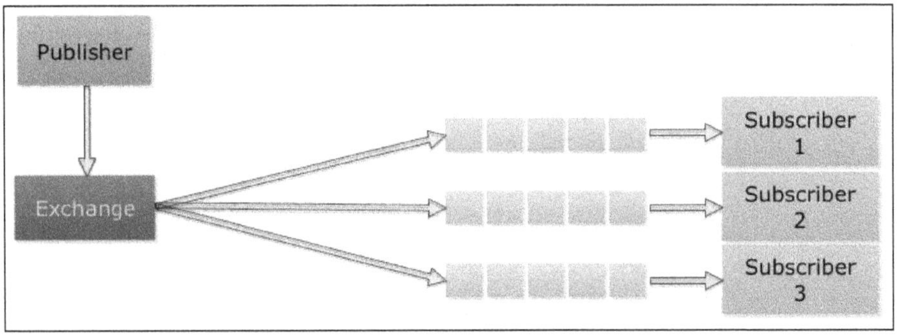

Bulk Message Architecture

Sender

The main responsibility of senders is publishing the message to the exchange that directly enqueues messages to the subscribed queues. To get this done, we should connect to the RabbitMQ Server first. Then, we declare the exchange with the fanout functionality, which gives us PubSub. Finally, we send our messages to subscribers with publishing the serialized message, as shown in the following code:

```
package com.collaboration.sender;

import java.io.IOException;

import com.collaboration.model.Message;
import com.rabbitmq.client.Channel;
import com.rabbitmq.client.Connection;
import com.rabbitmq.client.ConnectionFactory;


public class BulkSender {

  private final static String EXCHANGE_NAME =
  "mastering_rabbitmq_bulk";
```

```
    /**
     * @param args
     * @throws IOException
     */
    public static void main(String[] args) throws IOException {
      ConnectionFactory factory = new ConnectionFactory();
      factory.setHost("localhost");
      Connection connection = factory.newConnection();

      Channel channel = connection.createChannel();
      channel.exchangeDeclare(EXCHANGE_NAME, "fanout");

      Message msg = new Message();
      msg.setFrom("John");
      msg.setTo("Nicky");
      msg.setHeader("Hello World");
      msg.setContent("Hello World Again");

      for(int i = 0; i < 5; i++) {
        msg.setMsgNo(i + 1);
        channel.basicPublish(EXCHANGE_NAME, "", null,
        msg.toJSON().getBytes());
      }

      System.out.println("Message is sent: " + msg.toString());

      channel.close();
      connection.close();
    }
  }
```

Receiver

Receiver is called subscriber. Subscriber's main role is to subscribe to the publisher using its bound queue. Firstly, subscriber connects to the RabbitMQ Server. Next, subscriber declares the specific exchange with the fanout functionality. Then, subscriber waits for the incoming message in an infinite loop. Whenever a new message is published to the queues, subscriber fetches the serialized message and de-serialized message to the related `Message` object, as shown in the following subscriber implementation:

```
package com.collaboration.receiver;

import java.io.IOException;

import com.collaboration.model.Message;
import com.rabbitmq.client.Channel;
import com.rabbitmq.client.Connection;
import com.rabbitmq.client.ConnectionFactory;
import com.rabbitmq.client.ConsumerCancelledException;
import com.rabbitmq.client.QueueingConsumer;
import com.rabbitmq.client.ShutdownSignalException;

public class BulkReceiver {
  private final static String EXCHANGE_NAME =
  "mastering_rabbitmq_bulk";

  /**
   * @param args
   * @throws IOException
   * @throws ShutdownSignalException
   * @throws ConsumerCancelledException
   * @throws InterruptedException
   */
  public static void main(String[] args) throws IOException,
  ShutdownSignalException, ConsumerCancelledException,
  InterruptedException {
    ConnectionFactory factory = new ConnectionFactory();
    factory.setHost("localhost");

    Connection connection = factory.newConnection();
    Channel channel = connection.createChannel();

    channel.exchangeDeclare(EXCHANGE_NAME, "fanout");
    String queueName = channel.queueDeclare().getQueue();
    channel.queueBind(queueName, EXCHANGE_NAME, "");

    System.out.println("Waiting for the messages.........");

    QueueingConsumer consumer = new QueueingConsumer(channel);
    channel.basicConsume(queueName, true, consumer);

    while (true) {
```

```
            QueueingConsumer.Delivery delivery =
            consumer.nextDelivery();
            String msg = new String(delivery.getBody());

            System.out.println("Received: " +
            Message.fromJSON(msg).toString());
        }
    }
}
```

File message

In a collaboration application, we sometimes share a document, picture, or presentation to our group members. In RabbitMQ terms, we need to cover the file message inside the RabbitMQ Server to achieve the file message use case.

File message is not totally different with other message types. Moreover, you can use any other message type with file message. In file message, we should convert our files to the binary format.

Sender

As said earlier, sending file message is not different from other message types. We only need to make our files suit the messaging. Therefore, we should convert our files to the binary array. Then, we are ready to send our files into the messages, as shown in the following code:

```
package com.collaboration.sender;

import java.io.File;
import java.io.FileInputStream;
import java.io.IOException;
import java.io.InputStream;

import com.collaboration.model.FileMessage;
import com.rabbitmq.client.Channel;
import com.rabbitmq.client.Connection;
import com.rabbitmq.client.ConnectionFactory;

publicclass FileSender {
  privatefinalstatic String QUEUE_NAME =
  "mastering_rabbitmq_file";
```

```
/**
 * @param argv
 * @throws IOException
 */
publicstaticvoid main(String[] argv) throws IOException {
  ConnectionFactory factory = new ConnectionFactory();
  factory.setHost("localhost");
  Connection connection = factory.newConnection();

  Channel channel = connection.createChannel();
  channel.queueDeclare(QUEUE_NAME, false, false, false, null);

  FileMessage msg = new FileMessage();
  msg.setFrom("John");
  msg.setTo("Nicky");
  msg.setHeader("Hello World");
  msg.setFile_content(readBytesFromFile(new File("/Users/
  emrahayanoglu/Desktop/Desktop/RequiredComputers.txt")));

  for (int i = 0; i < 5; i++) {
    msg.setMsgNo(i + 1);
    channel.basicPublish("", QUEUE_NAME, null,
    msg.toJSON().getBytes());
  }

  System.out.println("File Message is sent: " + msg.toString());

  channel.close();
  connection.close();
}

/**
 * @param file
 * @return
 * @throws IOException
 */
publicstaticbyte[] readBytesFromFile(File file) throws
IOException {
  InputStream is = new FileInputStream(file);

  // Get the size of the file
  long length = file.length();
```

```
        // You cannot create an array using a long type.
        // It needs to be an int type.
        // Before converting to an int type, check
        // to ensure that file is not larger than Integer.MAX_VALUE.
        if (length > Integer.MAX_VALUE) {
          thrownew IOException("Could not completely read file "+
          file.getName() + " as it is too long (" + length+
          " bytes, max supported " + Integer.MAX_VALUE + ")");
        }

        // Create the byte array to hold the data
        byte[] bytes = newbyte[(int) length];

        // Read in the bytes
        int offset = 0;
        int numRead = 0;
        while (offset < bytes.length && (numRead = is.read(bytes,
        offset, bytes.length - offset)) >= 0) {
          offset += numRead;
        }

        // Ensure all the bytes have been read in
        if (offset < bytes.length) {
          thrownew IOException("Could not completely read file "+
          file.getName());
        }

        // Close the input stream and return bytes
        is.close();
        return bytes;
    }
}
```

Receiver

Considering that the sender of file messaging mostly seems like the other messaging types, receiver also behaves like the receiver of other message types. The difference between the receiver of file messaging and the other messaging types is de-serializing part of the messages. Whenever a new message comes to our queue, we should convert it from binary array to file, as shown in the following code example:

```
package com.collaboration.receiver;

import java.io.IOException;
```

```
import com.collaboration.model.FileMessage;
import com.rabbitmq.client.Channel;
import com.rabbitmq.client.Connection;
import com.rabbitmq.client.ConnectionFactory;
import com.rabbitmq.client.ConsumerCancelledException;
import com.rabbitmq.client.QueueingConsumer;
import com.rabbitmq.client.ShutdownSignalException;

public class FileReceiver {
  private final static String QUEUE_NAME =
  "mastering_rabbitmq_file";

  /**
   * @param argv
   * @throws IOException
   * @throws ShutdownSignalException
   * @throws ConsumerCancelledException
   * @throws InterruptedException
   */
  public static void main(String[] argv) throws IOException,
  ShutdownSignalException, ConsumerCancelledException,
  InterruptedException {
    ConnectionFactory factory = new ConnectionFactory();
    factory.setHost("localhost");

    Connection connection = factory.newConnection();
    Channel channel = connection.createChannel();

    channel.queueDeclare(QUEUE_NAME, false, false, false, null);

    System.out.println("Waiting for the messages........");

    QueueingConsumer consumer = new QueueingConsumer(channel);
    channel.basicConsume(QUEUE_NAME, true, consumer);

    while (true) {
      QueueingConsumer.Delivery delivery =
      consumer.nextDelivery();
      String msg = new String(delivery.getBody());

      System.out.println("Received: " +
      FileMessage.fromJSON(msg).toString());
    }
  }
}
```

RPC message

Remote Procedure Call (**RPC**) is a powerful technique for creating distributed, client-server based applications. Today, many distributed applications rely on the RPC. The main goal of RPC is to execute subroutine or procedure in the different server without knowing the details of the remote interaction.

RPC has a step-by-step functionality, as follows:

1. Client application calls the service.
2. Server executes the service.
3. Whenever server finishes the execution, client continues its execution.

We have lots of technology improvements over RPC technology. Also, we can call our services with the help of RabbitMQ, since it eases the communication between client and server in RPC architecture. Therefore, it is a more convenient way of using RabbitMQ between client and servers as seen in the following topics.

RPC client

RPC client's main aim is to request service calls and wait until it finishes executing. Therefore, we need two message queues between RPC client and RPC server. One is for sending requests to the RPC server and the other for upcoming finish replies for the RPC client.

Firstly, we should create our queue and send our request to the RPC server. Then, we should wait for the execution to finish, by means of listening to the queue for incoming messages, as shown in the following code example:

```
package com.collaboration.receiver;

import java.io.IOException;

import com.rabbitmq.client.AMQP.BasicProperties;
import com.rabbitmq.client.Channel;
import com.rabbitmq.client.Connection;
import com.rabbitmq.client.ConnectionFactory;
import com.rabbitmq.client.ConsumerCancelledException;
import com.rabbitmq.client.QueueingConsumer;
import com.rabbitmq.client.ShutdownSignalException;

publicclass RPCClient {
  privatefinalstatic String QUEUE_NAME = "mastering_rabbitmq_rpc";
```

```
    /**
     * @param args
     * @throws IOException
     * @throws ShutdownSignalException
     * @throws ConsumerCancelledException
     * @throws InterruptedException
     */
    publicstaticvoid main(String[] args) throws IOException,
    ShutdownSignalException, ConsumerCancelledException,
  InterruptedException {
      ConnectionFactory factory = new ConnectionFactory();
      factory.setHost("localhost");
      Connection connection = factory.newConnection();
      Channel channel = connection.createChannel();

      String replyQueueName = channel.queueDeclare().getQueue();
      QueueingConsumer consumer = new QueueingConsumer(channel);
      channel.basicConsume(replyQueueName, true, consumer);

      String response = null;
      String corrId = java.util.UUID.randomUUID().toString();

      BasicProperties props = new BasicProperties.Builder()
      .correlationId(corrId).replyTo(replyQueueName).build();

      String message = "10240000";

      channel.basicPublish("", QUEUE_NAME, props,
      message.getBytes());

      while (true) {
        QueueingConsumer.Delivery delivery =
        consumer.nextDelivery();
        if (delivery.getProperties().getCorrelationId().
        equals(corrId)) {
          response = new String(delivery.getBody());
          break;
        }
      }

      connection.close();
    }
  }
```

RPC server

RPC server's main role is to execute the given command, and after finishing the execution, notify the RPC client. Therefore, RPC server listens to the incoming message queue, and then executes the command. After finishing the execution, RPC server sends a message to the queue that is bound with RPC client to notify, as shown in the following code example:

```
package com.collaboration.sender;

import java.io.IOException;
import java.util.ArrayList;

import com.rabbitmq.client.AMQP.BasicProperties;
import com.rabbitmq.client.Channel;
import com.rabbitmq.client.Connection;
import com.rabbitmq.client.ConnectionFactory;
import com.rabbitmq.client.ConsumerCancelledException;
import com.rabbitmq.client.QueueingConsumer;
import com.rabbitmq.client.ShutdownSignalException;

public class RPCServer {
  private final static String QUEUE_NAME =
  "mastering_rabbitmq_rpc";

  /**
   * @param args
   * @throws IOException
   * @throws ShutdownSignalException
   * @throws ConsumerCancelledException
   * @throws InterruptedException
   */
  public static void main(String[] args) throws IOException,
  ShutdownSignalException, ConsumerCancelledException,
  InterruptedException {
    ConnectionFactory factory = new ConnectionFactory();
    factory.setHost("localhost");

    Connection connection = factory.newConnection();
    Channel channel = connection.createChannel();

    channel.queueDeclare(QUEUE_NAME, false, false, false, null);

    channel.basicQos(1);
```

```
    QueueingConsumer consumer = new QueueingConsumer(channel);
    channel.basicConsume(QUEUE_NAME, false, consumer);

    System.out.println(" [x] Waiting RPC requests");

    while (true) {
      QueueingConsumer.Delivery delivery =
      consumer.nextDelivery();

      BasicProperties props = delivery.getProperties();
      BasicProperties replyProps = new BasicProperties.Builder().
      correlationId(props.getCorrelationId()).build();

      String message = new String(delivery.getBody());
      int n = Integer.parseInt(message);

      System.out.println(" [.] nthPrimeList(" + message + ")");
      String response = "" + nthPrimeList(n);

      channel.basicPublish("", props.getReplyTo(), replyProps,
      response.getBytes());

      channel.basicAck(delivery.getEnvelope().getDeliveryTag(),
      false);
    }
  }

  /**
   * @param n
   * @return String
   */
  public static String nthPrimeList(int n) {
    ArrayList<Integer> primeList = new ArrayList<Integer>();
    for (int number = 2; number <= n; number++) {
      if (isPrime(number)) {
        primeList.add(number);
      }
    }
    return primeList.toString();
  }

  /**
   * @param number
```

```
    * @return boolean
    */
   public static boolean isPrime(int number) {
     for (int i = 2; i < number; i++) {
       if (number % i == 0) {
         return false; // number is divisible so its not prime
       }
     }
     return true; // number is prime now
   }
 }
```

Creating tasks – manual acknowledgment

Tasks are our main concern in our software systems nowadays. Processing documents, image processing, backup databases, and so on are the main examples of the tasks. As we discussed earlier, our main concern is to reply to requests in real-time. So, tasks are the main headaches of the systems.

Although tasks are problems for real-time systems, we should use RPC-like systems to solve the task problems, such as when some task handlers execute the incoming tasks and then reply to the task creators. Hence, task creator is not able to block it and execute its own work during the execution of the task.

RabbitMQ is the main part of the task queue systems. Task creator sends messages to the RabbitMQ queue, and Task handler listens to the queue and executes the dequeued task.

Task creator

Task creator's main responsibility is to create the related task along with sending the task message to the RabbitMQ queue. So, task creator just declares a queue and sends its task messages to the queue, as shown in the following code example:

```
package com.collaboration.sender;

import java.io.IOException;
import java.util.Calendar;
import java.util.Date;

import com.collaboration.model.Task;
import com.rabbitmq.client.Channel;
```

```
import com.rabbitmq.client.Connection;
import com.rabbitmq.client.ConnectionFactory;

publicclass TaskCreator {
  privatefinalstatic String QUEUE_NAME =
  "mastering_rabbitmq_task";

  /**
   * @param args
   * @throws IOException
   */
  publicstaticvoid main(String[] args) throws IOException {
    ConnectionFactory factory = new ConnectionFactory();
    factory.setHost("localhost");
    Connection connection = factory.newConnection();

    Channel channel = connection.createChannel();
    channel.queueDeclare(QUEUE_NAME, true, false, false, null);

    Calendar cal = Calendar.getInstance();
    cal.setTime(new Date());
    cal.add(Calendar.HOUR_OF_DAY, 2);

    Task task = new Task();
    task.setExpires(cal.getTime());
    task.setCommand("dd if=//dev//zero of=output.dat  bs=1024
    count=1024000");
    System.out.println(task.toJSON());

    for(int i = 0; i < 5; i++) {
      channel.basicPublish("", QUEUE_NAME, null,
      task.toJSON().getBytes());
      System.out.println("Task Request is sent: " +
      task.toString());
    }

    channel.close();
    connection.close();
  }
}
```

Task handler

Task handler simply waits for the upcoming tasks and executes the task that is dequeued from the RabbitMQ queue. The main difference between the simple sender and receiver application and the task creator and task handler application is to acknowledge manually. As you see in the following code example, after finishing the execution of the task, channel acknowledges manually, which notifies the RabbitMQ Server:

```
package com.collaboration.receiver;

import java.io.IOException;

import com.collaboration.model.Task;
import com.collaboration.utility.TaskRunner;
import com.rabbitmq.client.Channel;
import com.rabbitmq.client.Connection;
import com.rabbitmq.client.ConnectionFactory;
import com.rabbitmq.client.ConsumerCancelledException;
import com.rabbitmq.client.QueueingConsumer;
import com.rabbitmq.client.ShutdownSignalException;


public class TaskHandler {
  private final static String QUEUE_NAME =
  "mastering_rabbitmq_task";

  /**
   * @param argv
   * @throws IOException
   * @throws ShutdownSignalException
   * @throws ConsumerCancelledException
   * @throws InterruptedException
   */
  public static void main(String[] argv) throws IOException,
  ShutdownSignalException, ConsumerCancelledException,
  InterruptedException {
    ConnectionFactory factory = new ConnectionFactory();
    factory.setHost("localhost");

    Connection connection = factory.newConnection();
    Channel channel = connection.createChannel();
```

```
        channel.queueDeclare(QUEUE_NAME, true, false, false, null);

        System.out.println("Waiting for the tasks.........");

        QueueingConsumer consumer = new QueueingConsumer(channel);
        channel.basicConsume(QUEUE_NAME, false, consumer);

        while (true) {
          QueueingConsumer.Delivery delivery =
          consumer.nextDelivery();
          String msg = new String(delivery.getBody());

          System.out.println(msg);

          Task task = Task.fromJSON(msg);

          System.out.println("Received: " + task.toString());
          TaskRunner.runTask(task);
          System.out.println("Task is Done: " + task.toString());

          channel.basicAck(delivery.getEnvelope().getDeliveryTag(),
          false);
        }
      }
    }
```

Creating distributing tasks

Sometimes, it is not enough to achieve real-time processing and scalability with the single task handler. RabbitMQ comes to the rescue here. If a task handler executes a task more than once, a group of task handlers behave like a distributed style. Hence, we gain lots of time with the help of more task handlers.

Executing more than one task handler fetches task messages from the RabbitMQ queue with round robin scheduling. Round robin scheduling is one of the algorithms used for process and network scheduling in Computer Science. RabbitMQ distributes the task messages to the handlers with round robin style, as seen in the following task creator and task handler examples.

Task creator

The implementation of task creator code is as follows:

```
package com.collaboration.sender;

import java.io.IOException;
import java.util.Calendar;
import java.util.Date;

import com.collaboration.model.Task;
import com.rabbitmq.client.Channel;
import com.rabbitmq.client.Connection;
import com.rabbitmq.client.ConnectionFactory;

publicclass TaskCreator {
  privatefinalstatic String QUEUE_NAME =
  "mastering_rabbitmq_task";

  /**
   * @param args
   * @throws IOException
   */
  publicstaticvoid main(String[] args) throws IOException {
    ConnectionFactory factory = new ConnectionFactory();
    factory.setHost("localhost");
    Connection connection = factory.newConnection();

    Channel channel = connection.createChannel();
    channel.queueDeclare(QUEUE_NAME, true, false, false, null);

    Calendar cal = Calendar.getInstance();
    cal.setTime(new Date());
    cal.add(Calendar.HOUR_OF_DAY, 2);

    Task task = new Task();
    task.setExpires(cal.getTime());
    task.setCommand("dd if=//dev//zero of=output.dat  bs=1024
    count=1024000");
    System.out.println(task.toJSON());
```

```
        for(int i = 0; i < 5; i++) {
          channel.basicPublish("", "MyKey", null,
          task.toJSON().getBytes());
          System.out.println("Task Request is sent: " +
          task.toString());
        }

        channel.close();
        connection.close();
      }
    }
```

Task handler clients

The implementation of Task handler is as follows:

```
package com.collaboration.receiver;

import java.io.IOException;

import com.collaboration.model.Task;
import com.collaboration.utility.TaskRunner;
import com.rabbitmq.client.Channel;
import com.rabbitmq.client.Connection;
import com.rabbitmq.client.ConnectionFactory;
import com.rabbitmq.client.ConsumerCancelledException;
import com.rabbitmq.client.QueueingConsumer;
import com.rabbitmq.client.ShutdownSignalException;

public class DistributedTaskHandler {
  private final static String QUEUE_NAME =
  "mastering_rabbitmq_distributed_task";

  /**
   * @param argv
   * @throws IOException
   * @throws ShutdownSignalException
   * @throws ConsumerCancelledException
   * @throws InterruptedException
   */
  public static void main(String[] argv) throws IOException,
  ShutdownSignalException, ConsumerCancelledException,
  InterruptedException {
```

```
        ConnectionFactory factory = new ConnectionFactory();
        factory.setHost("localhost");

        Connection connection = factory.newConnection();
        Channel channel = connection.createChannel();

        channel.queueDeclare(QUEUE_NAME, true, false, false, null);

        System.out.println("Waiting for the messages.........");

        QueueingConsumer consumer = new QueueingConsumer(channel);
        channel.basicConsume(QUEUE_NAME, false, consumer);

        while (true) {
          QueueingConsumer.Delivery delivery =
          consumer.nextDelivery();
          String msg = new String(delivery.getBody());

          Task task = Task.fromJSON(msg);

          System.out.println("Received: " + task.toString());
          TaskRunner.runTask(task);
          System.out.println("Task is Done: " + task.toString());

          channel.basicAck(delivery.getEnvelope().getDeliveryTag(),
          false);
        }
      }
    }
```

Spring framework and RabbitMQ

Spring Framework is widely used in enterprise projects, and it is an open source application framework and inversion of control container for Java platform. **Inversion of Control** (**IoC**) provides consistency, that is, configuring and managing Java objects using reflection. The container for Spring Framework is responsible for creating these objects, calling their methods, and wiring them.

Spring Framework obtains the objects from the configuration file to initialize them, and is called **dependency injection**. Dependency injection is a well-known pattern that searches for the dependencies and injects the objects to the dependent objects with the help of constructor, properties, or factory methods.

Spring's amazing integration skills help libraries to integrate with systems more easily than ever. After this, it comes to RabbitMQ and Spring Framework. RabbitMQ client library can be injected to the dependent other systems easily with Spring Framework too. Moreover, Spring Framework's subproject called Spring AMQP aims to solve integration and injection problems of RabbitMQ, as discussed further in the following topic.

Spring AMQP

Message brokers are the main concern and integration part of software systems, and **AMQP** is the most popular messaging protocol that is backed by message brokers. Because of the importance of the AMQP, Spring community wanted to start a new project covering the AMQP integration within the Spring Framework. Then **Spring AMQP** was born, with these activities.

Spring AMQP provides a template as a high level abstraction to send and receive messages according to the Spring AMQP documentation. Spring AMQP gives us amazing classes to develop sender and receiver applications easily, and it provides two main classes:

- `RabbitTemplate` to send and receive messages
- `RabbitAdmin` to declare queues, exchanges, and bindings

The following case study examples show how to use Spring AMQP in developing messaging applications.

Single message

As we know from the Java RabbitMQ Client library examples, we need one queue to connect sender with receiver. To develop this kind of application in Spring AMQP, we should define the connection with the help of `rabbit:connection-factory` and `rabbit:template`.

We provide connection parameters such as hostname, port, username, and password information to `rabbit:connection-factory`. Then, we need to define queue with Rabbit Queue. Finally, we are ready to inject our connection and queue to the `rabbit:template` instance. Now, we are able to inject the RabbitMQ template to our senders and listeners. Listener container works as an asynchronous way in Spring AMQP, which is called `rabbit:listener-container`. Additionally, Spring AMQP provides message converters from JSON or XML, as shown in the following example configuration and code.

Spring config

A configuration for Spring can be specified as follows:

```
<?xml version="1.0" encoding="UTF-8"?>
<beans xmlns="http://www.springframework.org/schema/beans"
  xmlns:xsi="http://www.w3.org/2001/XMLSchema-instance"
  xmlns:int="http://www.springframework.org/schema/integration"
  xmlns:int-amqp="http://www.springframework.org/schema/
  integration/amqp"
  xmlns:rabbit="http://www.springframework.org/schema/rabbit"
  xmlns:context="http://www.springframework.org/schema/context"
  xmlns:int-stream="http://www.springframework.org/schema/
  integration/stream"
  xsi:schemaLocation="http://www.springframework.org/schema/
  integration/amqp
  http://www.springframework.org/schema/integration/amqp/spring-
  integration-amqp.xsd
    http://www.springframework.org/schema/integration
    http://www.springframework.org/schema/integration/spring-
    integration.xsd
    http://www.springframework.org/schema/integration/stream
    http://www.springframework.org/schema/integration/stream/
    spring-integration-stream.xsd
    http://www.springframework.org/schema/rabbit
    http://www.springframework.org/schema/rabbit/spring-rabbit.xsd
    http://www.springframework.org/schema/beans
    http://www.springframework.org/schema/beans/spring-beans-
    3.0.xsd
    http://www.springframework.org/schema/context
    http://www.springframework.org/schema/context/spring-context-
    3.0.xsd">

  <context:component-scan base-package="com.collaboration"/>

  <bean id="messageListener" class="com.collaboration.receiver.
  Receiver"/>

  <bean id="sender" class="com.collaboration.sender.Sender"/>

  <bean id="messageConverter" class="org.springframework.amqp.
  support.converter.JsonMessageConverter"/>

  <rabbit:queue id="masteringQueue" name="mastering.rabbitmq"/>
```

```
    <rabbit:connection-factory id="rabbitConnectionFactory"
    username="guest" password="guest" host="localhost" port="5672"/>

    <rabbit:template id="rabbitTemplate" connection-
    factory="rabbitConnectionFactory"
    queue="masteringQueue" message-converter="messageConverter"/>

    <rabbit:admin id="admin" connection-
    factory="rabbitConnectionFactory"/>

    <rabbit:listener-container connection-factory=
    "rabbitConnectionFactory"  message-converter="messageConverter">
      <rabbit:listener ref="messageListener"
      queues="masteringQueue"/>
    </rabbit:listener-container>

</beans>
```

Sender

Sender code is as follows:

```
package com.collaboration.sender;

import org.springframework.amqp.core.AmqpTemplate;
import org.springframework.beans.factory.annotation.Autowired;

/**
 * @author Emrah Ayanoglu
 *
 */
public class Sender {
  @Autowired
  private volatile AmqpTemplate amqpTemplate;

  /**
   * Sends new Message using AmqpTemplate
   */
  public void sendMessage(){
    amqpTemplate.convertAndSend("Hello World");
  }
}
```

Receiver

Receiver code is as follows:

```
package com.collaboration.receiver;

import org.springframework.amqp.core.Message;
import org.springframework.amqp.rabbit.core.
ChannelAwareMessageListener;

import com.rabbitmq.client.Channel;

/**
 * @author Emrah Ayanoglu
 *
 */
public class Receiver implements ChannelAwareMessageListener {

  /* (non-Javadoc)
   * @see org.springframework.amqp.rabbit.core.
  ChannelAwareMessageListener#onMessage(org.springframework.
  amqp.core.Message, com.rabbitmq.client.Channel)
   */
  public void onMessage(Message message, Channel channel) throws
  Exception {
    System.out.println("A message is received : Receiver");
    String msgBody = new String(message.getBody());
    System.out.println("Message: " + msgBody);
  }

}
```

PubSub messages

Publish and Subscribe style messaging needs exchange and connected queues to the provided exchange. In Spring AMQP, we have another opportunity to define pubsub exchanges, which is `rabbit:fanout-exchange`. We should inject connected queues to the `rabbit:fanout-exchange` with `rabbit:bindings`, as shown in the following configuration example.

Spring config

A configuration for spring can be specified as follows:

```
<?xml version="1.0" encoding="UTF-8"?>
<beans xmlns="http://www.springframework.org/schema/beans"
  xmlns:xsi="http://www.w3.org/2001/XMLSchema-instance"
  xmlns:int="http://www.springframework.org/schema/integration"
  xmlns:int-amqp="http://www.springframework.org/schema/
  integration/amqp"
  xmlns:rabbit="http://www.springframework.org/schema/rabbit"
  xmlns:context="http://www.springframework.org/schema/context"
  xmlns:int-stream="http://www.springframework.org/schema/
  integration/stream"
  xsi:schemaLocation="http://www.springframework.org/schema/
  integration/amqp
  http://www.springframework.org/schema/integration/amqp/spring-
  integration-amqp.xsd
    http://www.springframework.org/schema/integration
    http://www.springframework.org/schema/integration/spring-
    integration.xsd
    http://www.springframework.org/schema/integration/stream
    http://www.springframework.org/schema/integration/stream/
    spring-integration-stream.xsd
    http://www.springframework.org/schema/rabbit
    http://www.springframework.org/schema/rabbit/spring-rabbit.xsd
    http://www.springframework.org/schema/beans
    http://www.springframework.org/schema/beans/spring-beans-
    3.0.xsd
    http://www.springframework.org/schema/context
    http://www.springframework.org/schema/context/spring-context-
    3.0.xsd">

  <context:component-scan base-package="com.collaboration"/>

  <bean id="message1Listener" class=
  "com.collaboration.receiver.Receiver"/>
  <bean id="message2Listener" class=
  "com.collaboration.receiver.Receiver2"/>
  <bean id="message3Listener" class=
  "com.collaboration.receiver.Receiver3"/>
```

```
  <bean id="sender" class="com.collaboration.sender.Sender"/>

  <bean id="messageConverter" class="org.springframework.amqp.
  support.converter.JsonMessageConverter"/>

  <rabbit:queue id="mastering1Queue" name="mastering1.rabbitmq"/>
  <rabbit:queue id="mastering2Queue" name="mastering2.rabbitmq"/>
  <rabbit:queue id="mastering3Queue" name="mastering3.rabbitmq"/>

  <rabbit:fanout-exchange name="broadcast.responses"
  xmlns="http://www.springframework.org/schema/rabbit">
  <rabbit:bindings>
  <rabbit:binding queue="mastering1Queue"/>
  <rabbit:binding queue="mastering2Queue"/>
  <rabbit:binding queue="mastering3Queue"/>
  </rabbit:bindings>
  </rabbit:fanout-exchange>

  <rabbit:connection-factory id="rabbitConnectionFactory"
  username="guest" password="guest" host="localhost" port="5672"/>

  <rabbit:template id="rabbitTemplate" connection-
  factory="rabbitConnectionFactory" exchange="broadcast.responses"
  message-converter="messageConverter"/>

  <rabbit:admin id="admin" connection-factory=
  "rabbitConnectionFactory"/>

  <rabbit:listener-container connection-factory=
  "rabbitConnectionFactory"  message-converter="messageConverter">
    <rabbit:listener ref="message1Listener" queues=
    "mastering1Queue"/>
    <rabbit:listener ref="message2Listener" queues=
    "mastering2Queue"/>
    <rabbit:listener ref="message3Listener" queues=
    "mastering3Queue"/>
  </rabbit:listener-container>

</beans>
```

Private messages – routing

In private messaging, that is, routed messaging in message broker terms, we use topic exchange in RabbitMQ. Spring AMQP gives us another awesome template for topic exchanges, which is called **topic-exchange**. Whenever binding queues to the topic-exchange, we need to specify the pattern of each queue. Whenever a new message is received from the topic exchange, it is delivered to the queues with respect to the patterns. You can find the details of the topic-exchange in the following example.

Spring config

A configuration for spring can be specified as follows:

```
<?xml version="1.0" encoding="UTF-8"?>
<beans xmlns="http://www.springframework.org/schema/beans"
  xmlns:xsi="http://www.w3.org/2001/XMLSchema-instance"
  xmlns:int="http://www.springframework.org/schema/integration"
  xmlns:int-amqp="http://www.springframework.org/schema/
  integration/amqp"
  xmlns:rabbit="http://www.springframework.org/schema/rabbit"
  xmlns:context="http://www.springframework.org/schema/context"
  xmlns:int-stream="http://www.springframework.org/schema/
  integration/stream"
  xsi:schemaLocation="http://www.springframework.org/schema/
  integration/amqp
  http://www.springframework.org/schema/integration/amqp/spring-
  integration-amqp.xsd
    http://www.springframework.org/schema/integration
    http://www.springframework.org/schema/integration/spring-
    integration.xsd
    http://www.springframework.org/schema/integration/stream
    http://www.springframework.org/schema/integration/stream/
    spring-integration-stream.xsd
    http://www.springframework.org/schema/rabbit
    http://www.springframework.org/schema/rabbit/spring-rabbit.xsd
    http://www.springframework.org/schema/beans
    http://www.springframework.org/schema/beans/spring-beans-
    3.0.xsd
    http://www.springframework.org/schema/context
    http://www.springframework.org/schema/context/spring-context-
    3.0.xsd">
  <context:component-scan base-package="com.collaboration"/>
```

```
    <bean id="message1Listener" class=
    "com.collaboration.receiver.Receiver"/>
    <bean id="message2Listener" class=
    "com.collaboration.receiver.Receiver2"/>
    <bean id="message3Listener" class=
    "com.collaboration.receiver.Receiver3"/>

    <bean id="sender" class="com.collaboration.sender.
    SenderWithRoutingKey"/>

    <bean id="messageConverter" class="org.springframework.amqp.
    support.converter.JsonMessageConverter"/>
    <rabbit:queue id="mastering1Queue" name="mastering1.rabbitmq"/>
    <rabbit:queue id="mastering2Queue" name="mastering2.rabbitmq"/>
    <rabbit:queue id="mastering3Queue" name="mastering3.rabbitmq"/>

    <topic-exchange name="routing.responses"
    xmlns="http://www.springframework.org/schema/rabbit">
    <bindings>
    <binding queue="mastering1Queue" pattern="private"/>
    <binding queue="mastering2Queue" pattern="group"/>
    <binding queue="mastering3Queue" pattern="all"/>
    </bindings>
    </topic-exchange>

    <rabbit:connection-factory id="rabbitConnectionFactory"
    username="guest" password="guest" host="localhost" port="5672"/>

    <rabbit:template id="rabbitTemplate" connection-factory=
    "rabbitConnectionFactory" exchange="routing.responses" message-
    converter="messageConverter"/>

    <rabbit:admin id="admin" connection-factory=
    "rabbitConnectionFactory"/>

    <rabbit:listener-container connection-factory=
    "rabbitConnectionFactory" message-converter="messageConverter">
    <rabbit:listener ref="message1Listener" queues=
    "mastering1Queue"/>
      <rabbit:listener ref="message2Listener" queues=
      "mastering2Queue"/>
      <rabbit:listener ref="message3Listener" queues=
      "mastering3Queue"/>
    </rabbit:listener-container>

</beans>
```

Summary

In this chapter, we learned how to develop RabbitMQ client using the Java platform and Java programming language. RabbitMQ officially provides us with a Java Client Library, and the library provides all of the functionality of AMQP within itself. Library provides both synchronous and asynchronous listeners. We shared the basics of developing the clients and also learned how to develop messaging facilities of the collaboration application using the RabbitMQ Server.

In the second part, we dived into the Spring integration with Spring's new project, Spring AMQP. Spring AMQP simplifies the developing clients by providing the template classes for the RabbitMQ Server.

The next chapter will cover developing clients using the C# and .Net Framework.

10

Ruby Client Programming

In this chapter, we will explore the most harmonious combination of client-broker there is (in my opinion), Ruby and RabbitMQ. You will quickly feel as if it's a match made in heaven.

You will learn how to build a real-life data science pipeline using Lambda Architecture, a worker fabric, and we will introduced to both Bunny—the de-facto RabbitMQ library for Ruby and Sneakers—and my own high-level performance background processing job library for Ruby.

In this chapter, you will learn how to the following:

- Use Bunny to implement each and every messaging semantics in Ruby
- Explore Lambda architecture and understand why this is the new way forward in a world filled with Big data
- Understand the differences between Big, Medium, and Small data
- See how to build a solution that would really hold such an architecture
- Implement the solution with Sneakers and build an `aggregator` type worker—an `ip2location` type worker; you will also understand how to go much further than this
- Explore a bit more of the `Sneakers` library, and see why sometimes there's way more to do in production other than just punch out code

Case study

For this study, we'll choose the world of Real Time Analytics. You will learn about the challenges in real-time data, and how they can be solved with scalability in mind. But first, let's talk about analytics, Small data, Medium data, and Big data.

Small data

Small data is when you have enough data to process, but fits into a single machine. Any processing and analysis you want to run can finish in a reasonable amount of time. Of course term "reasonable" may refer to different amount of time depending on your business needs.

For example, if we want to generate a daily report, it is reasonable to assume that we're still okay with it taking around 30 minutes to complete. This is because we will still have 47 other tries to make it happen (we have 48 half hours within a 24 hour day).

However, it is unreasonable to agree to a job or series of interdependent jobs that take more than, or very close to, 24 hours. In this case, you will seek to scale out of your single machine or workstation and start thinking about Medium data. Having said that, Small data is fun as it usually lets you get very high latencies. You can store all data in RAM and have any aggregation or analysis finish in milliseconds.

We'll talk about Medium data last, you'll soon see why.

Big data

Big data is when the amount of data you have can only fit on a cluster of distributed and coordinated machines. We refer to big data, when the amount of storage or processing power simply can't be satisfied with a single machine. We also refer to big data whenever the amount of time, or resources of jobs in any other setting (say, on your own machine) is so big that jobs would simply fail, let alone be slow to complete.

Big data is not easy to solve; however, today it is mostly solved. With frameworks such as **Hadoop** and **Spark** and well-mannered distributions such as **Cloudera Hadoop** (**CDH**), the monster that is **Hadoop DevOps** is easily, or at least somewhat, well-tamed.

The turnaround time with big data is slow. The development of jobs, the feedback cycle, and actual job runs are slow. We, as a community, improved this with Spark and **Hive** and **Pig** and **Cascalog**. However, it is still an *unnatural* development workflow.

If you have a Big data problem, you mostly also have a real-time problem as the two cannot live side by side. For this manner, we will introduce the Lambda architecture, which is an emerging architecture to get great Big data pipeline performance as well as real-time pipeline performance happening in the same time. Lambda architecture is a generic, scalable, and fault-tolerant data processing, where it takes advantage of both batch and stream processing methods. Lambda architecture consists of three layers: the layer, and serving layers.

The batch layer is designed to guarantee perfect accuracy by being able to process all available data when generating views. The speed layer is designed to process data streams in real-time and without the requirements of fix ups or completeness. The speed layer sacrifices throughput as it aims to minimize latency by providing real-time views into the most recent data. The serving layer responds to queries using both stream and batch results.

Either way, this world has matured. Best practices are there for you to explore, and you're probably going to be in good hands.

Medium data

As of today (somewhere around March 2015, as of the writing of this book), the tech world is only starting to realize there's another world of data living between Small and Big data.

The big problem is now converging to be Medium data. It is that state of limbo when your company generates too much of Big data in order to compute on a single machine, or a single, very costly and strong machine, and yet, it is too small to justify the overhead and funding for a full-on Hadoop cluster.

Things such as AWS EMR, which is a Hadoop-on-demand, were put in place to tackle this kind of scenario in terms of cost, but then you're still left with the unhappiness of the slow development experience and the job run turn-around feedback.

What's more, you're left with no answer for real-time data, and you find yourself trying a much optimized **PostgreSQL** database and make it perform to bootstrap for this mission. Otherwise, you may acknowledge that you don't have the tech chops of a DBA; if so, then try a huge **MongoDB** cluster (you'll need at least six machines, three shards, each with master slave); people arguably say that it will crash and lose data on you.

Solving all data problems

The smart thing to do is not to address Small, Big, or Medium data problems, but to try to go our own separate way. We will see how to implement a solution for an analytics engine that would precompute everything rather than push data to a database, or a Hadoop cluster as we will dissect later.

We will see how RabbitMQ and the Ruby ecosystem lets you build a Lambda architecture style solution with almost no effort. You can carry over this kind of solution to any startup or company you're part of and it will always, always work.

But first, let's meet Bunny.

Bunny and Ruby

Bunny is the most used AMQP/RabbitMQ library within the Ruby ecosystem. Bunny 0.9+ supports all RabbitMQ 3.x features. Moreover, the latest Bunny 0.9+ is designed to make use of concurrency. On Ruby VMs that provide thread parallelism, this means taking advantage of multiple cores and CPUs.

Installing Ruby

If you already have a modern Ruby installed (Ruby 1.9.x or 2.x), feel free to skip this part. If you don't have Ruby installed, follow through this simple explanation:

Linux

On Linux, you can use the `apt-get` command:

```
$ sudo apt-get install ruby
```

This will probably bring in a Ruby 1.9.3 installation, and if you're lucky and have a recent Linux distribution, a Ruby 2.0.0 (as of this writing, Ruby 2.2.0 exists, but I don't expect it to be streamlined into every Linux distribution).

To verify this, try the following commands, which will tell you about your Ruby and **Rubygems** versions (the Ruby's dependency manager):

```
$ ruby -v
ruby 1.9.3p392 (2013-02-22 revision 39386) [x86_64-darwin12.2.1]
$ gem -v
1.8.23
```

Windows

On Windows, I recommend using a one-click install, one of which exists here:

`http://rubyinstaller.org.`

Follow the wizard and then verify the installation in a Windows CMD window (your output may vary, but take note of the versions):

```
c:\>ruby -v
ruby 2.2.0p0 (2014-12-25 revision 49005)
c:\>gem -v
1.8.23
```

OSX/Mac

On a Mac, I adequately install everything with **Homebrew**. You'll do yourself a big favor if you start using it and start installing development software with it. You can install both Ruby and RabbitMQ with it.

To install Homebrew, visit `http://brew.sh/` and follow the instructions at the bottom of the page where we refer to brew.sh URL.

After you've got Homebrew provisioned on your machine, you can open a terminal and type in the following:

```
$ brew install ruby
```

And here, again, you should verify that everything works as expected; run the following commands in your terminal:

```
➜ ~ ruby -v
ruby 2.2.0p0 (2014-12-25 revision 49005) [x86_64-darwin13]
➜ ~ gem -v
2.4.5
```

Rbenv

There's another option for installing Ruby across all the platforms using a tool called **Rbenv**; I personally use this. Rbenv will let you jump across Ruby versions easily and install and test out new Ruby distributions with the help of Rbenv's plugin called `ruby-build`.

Installing Rbenv is a bit out of the scope of this chapter, but I couldn't leave it out as it is the Swiss army knife of every Rubyist. If you're feeling capable and adventurous, feel free to follow the instructions for installing Rbenv at the RbenvGithub repository at the following link: `https://github.com/sstephenson/rbenv`.

Installing Bunny

Let's move on to installing and verifying our Bunny installation. Firstly, let's use Rubygems to install it:

```
$ gem install bunny
Fetching: bunny-1.7.0.gem (100%)
Successfully installed bunny-1.7.0
1 gem installed
```

```
Installing ri documentation for bunny-1.7.0...
Installing RDoc documentation for bunny-1.7.0...
```

Then, let's verify it by opening an interactive Ruby session using `irb`, a command line Ruby tool:

```
$ irb
irb(main):001:0> require 'bunny'
=>true
irb(main):002:0>Bunny.new
=> #<Bunny::Session:70143003791640 guest@127.0.0.1:5672, vhost=/,
hosts=[127.0.0.1]>
```

We managed to acquire Bunny; that is, we can now use Bunny anywhere in our Ruby code and then start a new Bunny session! However, `Bunny.new` command will throw an error if there is no RabbitMQ running locally.

Using Bunny

Let's continue by building a sanity-level consumer and producer just to test things out.

Bunny producer

We will first have a taste of what a log aggregator producer looks like:

```
require "bunny"

conn = Bunny.new
conn.start

channel = conn.create_channel
queue = channel.queue("clicks")

channel.default_exchange.publish('{ "message":"hello" }', :routing_key
=> queue.name)
puts "* sent!"

conn.close
```

We start inquiring about Bunny so that we have access to the Bunny API. We then initialize a Bunny instance, start a connection for the purpose of getting a reach at an AMQP channel, and then through the channel, we declare a queue and get access to an exchange.

We use the default exchange for simplicity, which was accessed through the channel we just got. We publish through a routing key that incidentally (or not) has the same name as the queue in order to autoroute the message to that queue.

We send a message that looks like JSON, but this is just a hint at what's about to come; for all intents and purposes, you can just send a plain string.

Let's make sure this runs:

```
$ rubyproducer.rb
W, [2015-03-06T18:33:11.838609 #34238]  WARN --
#<Bunny::Session:70145632515880 guest@127.0.0.1:5672, vhost=/,
hosts=[127.0.0.1]>: Could not establish TCP connection to
127.0.0.1:5672:
/Users/dotan/.rbenv/versions/1.9.3-p392/lib/ruby/gems/1.9.1
/gems/bunny-1.7.0/lib/bunny/session.rb:302:in `rescue in start':
Could not establish TCP connection to any of the configured hosts
(Bunny::TCPConnectionFailedForAllHosts)
  from /Users/dotan/.rbenv/versions/1.9.3-p392/lib/ruby/gems/1.9.1
  /gems/bunny-1.7.0/lib/bunny/session.rb:264:in `start'
  from producer.rb:4:in `<main>'
```

Bam! It failed. I like to do this; you'll see this coming in the Python chapter as well. I like the fact that the first encounter with a library should include errors. You should be able to feel the limits of the context that you're working with, and it will make learning a stronger and better experience just by bumping your head against an error here and there.

In this case, we can easily glean that we can't connect to a broker. This is probably because a broker isn't alive on my machine. If this happens to you too, start the broker and continue.

Just again:

```
$ rubyproducer.rb
* sent!
```

Great, let's verify that we have a message in store:

```
$ rabbitmqctllist_queues
Listing queues ...
celery 0
clicks 1
downloads 0
foobar 1
```

```
logs 0
test_stress 0
testqueue 0
usages_ 0
webscraping 0
```

We have one message in the clicks queue right here. Awesome!

Bunny consumer

Let's make in the same vein a simple sanity Bunny consumer:

```
require "bunny"
require 'json'

conn = Bunny.new
conn.start

channel = conn.create_channel
queue = channel.queue("clicks")

puts "* Consumer started."
queue.subscribe(:block => true) do |delivery_info, properties,
msg|
puts "got message: #{ JSON.parse(msg)["message"] }"
end
```

Let's go over the code. First, we, again, require Bunny, which as you might recall, lets us have a go at the Bunny's API. We then go through the same dance of getting a Bunny instance, which lets us create a channel by establishing a connection.

Through the channel, we declare and bind to a queue, the same "clicks" queue that we have used before in the producer of course and then we do something new—we use the queue that we have obtained in order to subscribe.

Subscription in Bunny is very powerful. Powerful in the sense that Bunny will hand out a set of objects to you, expect you to do something with it, and that's it. Bunny hides away all of the ugly details from you using the channel by blocking through the connection, working with the AMQP protocol, and hiding the execution model; does each block run on a thread of its own or on a single thread?

With Bunny, you shouldn't really care, as you are presented with a simple Ruby-like workflow, and this is why it is awesome.

Back to the code, you're telling Bunny that you want to block. This means, you want Bunny to keep doing the message loop endlessly (well, at least as long there's a good connection open to your broker). It will then give out a `delivery_info` properties, and message objects. All of these are part of the AMQP model, and trivially, you should probably care about the message.

Once you get a message, you can process it in your processing block. Here, we just spiced it up with trying and parsing the data within the message. As you recall, the data is a simple JSON object that is serialized before getting pushed into the queue.

Here, we deserialize it using the standard Ruby JSON library (see our require 'JSON' beforehand), and we pull the `message` property from it; job well done!

Let's start the consumer in your terminal type:

```
$ rubyconsumer.rb
* Consumer started.
got message: hello
```

Exploring the AMQP model with Bunny

Let's continue on with Bunny by exploring a bit more of the AMQP model. As it stands, the AMQP model is so powerful that it will let you abstract out a lot of messaging architectures pretty easily (I personally think that only ZeroMQ comes close to it in the same way as being a building block of many other concepts within the world of messaging).

Let's continue case by case.

Workers

Let's touch the concept of background queues and workers lightly. I say lightly because I plan to introduce a popular and production grade background jobs library called Sneakers - by yours truly, later on in this chapter.

The concept of workers is very close to what we've seen so far. We will only slightly change the semantics of our producer and call it a manager (`manager.rb`):

```
require "bunny"

conn = Bunny.new
conn.start

channel = conn.create_channel
```

```
queue = channel.queue("jobs", :durable => true)

queue.publish("do this!", :persistent => true)
puts "* Sent."

conn.close
```

What's changed here is that we're now using a full-on queue to publish. We initialize Bunny as usual and get a channel. Through the channel, we will create a durable queue, which will be used to hold our jobs.

We select a durable queue because, as expected of a job queue, it should exist when the broker crashes, and we can't afford to be in a situation where jobs are lost.

Continuing on, we then publish a persistent message. A note on publishing—RabbitMQ will never promise to persist, but it will do its next best effort to do so. If you really need a high degree of assurance, you should use Publish Confirms that are available to you form RabbitMQ. However, there's a trade-off of durability and performance.

Let's run our manager now:

```
$ rubymanager.rb
* Sent.
```

We will now have a message in the `"jobs"` durable queue. Let's continue to build our actual worker:

```
require "bunny"

conn = Bunny.new
conn.start

channel = conn.create_channel
queue = channel.queue("jobs", :durable => true)

channel.prefetch(1)

queue.subscribe(:manual_ack => true, :block => true) do |delivery_
info, properties, msg|
work_for = rand
sleep(work_for)
puts "* Done with: '#{msg}' in #{work_for}sec"
channel.ack(delivery_info.delivery_tag)
end
```

We start out again by requiring Bunny and making a connection and a channel as usual. We create a durable `"jobs"` queue. We then define a prefetch level. A prefetch level or window is the amount of messages our worker should fetch from the broker in each trip. For example, if we say "50", we are telling the broker that our worker wants to handle 50 messages in a bulk each time, and the broker will push down a batch of 50 messages on each communication trip.

This means less overhead and less "nagging" from our client, which means greater throughput. Our worker will be doing more actual work instead of more communication work per job.

With this, of course, there are trade-offs. The trade-off here is that of latency versus throughput. If we select a bigger batch, it will mean that a job X within the batch will take more time to be reported as "complete" by the worker.

Usually, the rule of thumb is that if the character of a job is different to the worker from job to job, for example—let's say one job is to "send an e-mail" and one job is to "calculate 1+1", and the worker can never expect which job comes next—fix the prefetch level at 1.

If the jobs are always similar, fix the prefetch level at a number which you feel comfortable with after trial-and-error: 20, 50, or even 100. Choose anything that works better for you in terms of latency and throughput.

Back to the code, we subscribe as before, but just with a new `manual_ack` flag set to `true`. This means we want each worker to manually sign off the job as done when it is actually done.

Within the worker block, we sleep for a random amount of time, and then when we're ready, we report the job as done by specifying `channel.ack` and the required `delivery_tag` flag.

This concludes our worker code; let's see it in practice. In one terminal push a few messages on, as follows:

```
$ watch ruby manager.rb
Every 2.0s: ruby manager.rb
Fri Mar  6 19:27:05 2015
* Sent.
```

The `watch` command is a cool Unix trick (on OSX, run '`brew install watch`') that you can use in order to repeat a command with a specified interval. Here, we will run '`ruby manager.rb`' every two seconds, which will in turn push the same message to our broker every two seconds—a perfect solution for a dummy manager.

Now, in a few other terminals, let's run a couple of workers, and let them compete for jobs and pick off and sign off jobs from the job queue:

Here's one:

```
$ rubyworker.rb
* Done with: 'do this!' in 0.686646707588058sec
* Done with: 'do this!' in 0.5178492461693734sec
* Done with: 'do this!' in 0.38852506368321216sec
```

And here's another:

```
$ rubyworker.rb
* Done with: 'do this!' in 0.8007668590674376sec
* Done with: 'do this!' in 0.39105215729062026sec
* Done with: 'do this!' in 0.20513470050165317sec
* Done with: 'do this!' in 0.7273595184529634sec
* Done with: 'do this!' in 0.23883835990736701sec
```

Workers keep churning until the queue is depleted and all jobs are performed. This completes our overview of manager/worker type of messaging semantics. Remember the key points here—`manual_ack` flag and prefetch.

Publish – subscribe

Another type of messaging semantics which is very popular within Mobiles today is pub/sub, publish-subscribe, or "push" as it's called in the Mobile domain (Google—Android: GCM, Apple: APNS).

Bunny and RabbitMQ lets you effortlessly model pub/sub. Let's see how this happens by starting off with a publisher:

```
require "bunny"

conn = Bunny.new
conn.start

channel = conn.create_channel
exchange = channel.fanout("push")

exchange.publish("testing: 1,2,3.")
puts "* Sent."

conn.close
```

Let's skip to the meat. We're creating a new type of object here on demand. We've created a `fanout` exchange. In general, this kind of exchange in RabbitMQ and AMQP is used to implement a pub/sub type of semantics.

To refresh our memory, when we push messages to a `fanout` exchange, they will appear on every bound queue.

We will need the subscribers to be able to hook into this stream of messages from this kind of exchange easily. This will mean that our subscribers will have to be created and killed without much drama. In other words, they will be transient, and we will create and bind them to the exchange in a transient way. Let's see the code for the subscriber now:

```
require "bunny"

conn = Bunny.new
conn.start

channel = conn.create_channel
exchange = channel.fanout("push")
queue = channel.queue('', :exclusive => true)

queue.bind(exchange)

sub_id = rand(1000)
puts "Subscribed to topic 'push'."
queue.subscribe(:block => true) do |delivery_info, properties,
msg|
puts "Subscriber(#{sub_id}): got #{msg}."
end

channel.close
conn.close
```

So, here again, as we are skipping the connection and channel creation ceremonies, we are presented with a new kind of construct, as in the same case of the publisher. We create and bind to a `fanout` exchange, but now we make a special kind of queue.

We choose not to name the queue and make it exclusive. This will create the effect of building a kind of an anonymous queue that's held by our process only. Given that we want to run several subscribers (remember that we are building a pub/sub hub), this fits us perfectly.

Let's give this thing a run and see what we're getting. We will again use the most useful Unix '`watch`' command for messages to keep being pushed by our publisher.

So, in one terminal, let's run the following:

```
$ watch ruby publisher.rb
```

This will immediately spawn into a process that keeps running the same command, by a default of 2.0 seconds.

In multiple terminals, let's run the following command:

```
$ rubysubscriber.rb
```

And the outputs will soon appear as the following on Terminal 1:

```
$ rubysubscriber.rb
Subscribed to topic 'push'.
Subscriber(131): got testing: 1,2,3..
Subscriber(131): got testing: 1,2,3..
Subscriber(131): got testing: 1,2,3..
Subscriber(131): got testing: 1,2,3..
```

And on the other terminal, Terminal 2, we will find this:

```
$ rubysubscriber.rb
Subscribed to topic 'push'.
Subscriber(908): got testing: 1,2,3..
Subscriber(908): got testing: 1,2,3..
Subscriber(908): got testing: 1,2,3..
Subscriber(908): got testing: 1,2,3..
```

As far as we're concerned, this works; we have two different subscribers that output the same kind of message previously pushed by a publisher!

Routing

We'll move onto another pillar of AMQP—**routing**. Let's see how we can intelligently route messages to different recipients. In this case, we will model an e-mail service, where every process has its own inbox. This is a bit similar to modeling an RPC mechanism (which we've explored earlier):

```
require "bunny"

conn = Bunny.new
conn.start
```

```
channel = conn.create_channel
exchange = channel.direct("mailbox")

puts "Sending to #{ARGV[0]}"
exchange.publish("testing: 1,2,3", :routing_key => ARGV[0])

conn.close
```

Here, we are creating a direct exchange. This kind of exchange will help us model a routing semantics at the top of AMQP. We will publish it with a route, or address, that we get from the command line with `ARGV[0]`; the rest is just gluing things up so that the publisher will push to this address. Let's run it:

```
$ruby sender.rbamsterdam
* Sending to amsterdam.
```

It works! Now, let's set up the more complex yet simple enough recipient, which will listen on a specified address or route:

```
require "bunny"

conn = Bunny.new
conn.start

channel = conn.create_channel
exchange = channel.direct("mailbox")
queue    = channel.queue("", :exclusive => true)

puts "* Accepting messages on address: #{ARGV[0]}"
queue.bind(exchange, :routing_key => ARGV[0])

queue.subscribe(:block => true) do |delivery_info, properties,
msg|
puts "* Got #{msg}"
end
```

Let's walk through this code. First, after getting a connection and a channel, we again set up a direct exchange, which will allow us to bind onto a specific route.

We then, similar to what we did with the pub/sub model, create an exclusive and anonymous queue and bind to this exchange. When we bind—this is the important part—we bind to the desired route as well.

Next up is our by-now familiar piece of subscription code, which takes a message and prints it.

Let's see how everything runs. Start three terminals, keep one for the sender, and run two of the recipients, each on a different address:

```
$ rubyrecipient.rbparis
* Accepting messages on address: paris
```

And for the second recipient, do as follows:

```
$ rubyrecipient.rbamsterdam
* Accepting messages on address: amsterdam
```

Now, let's run our sender; each time we will deliver to a different address. Keep a good eye on the couple of other terminals, and you'll see them light up in each turn:

```
$ rubysender.rbparis
* Sending to paris.
rubysender.rbamsterdam
* Sending to amsterdam.
```

And now, the desired output on each of the terminals is shown here:

```
* Accepting messages on address: paris
* Got testing: 1,2,3
* Accepting messages on address: amsterdam
   * Got testing: 1,2,3
```

Nice! Routing is a powerful concept—so powerful that a lot of products have been implemented easily just by reusing RabbitMQ's routing engine. Take a look at a product called **Sensu**, for example, a widely deployed, highly scalable cloud monitoring product, which is considered an evolution of **Zabbix** and **Nagios.Sensu**. It connects the output from "`check`" scripts with "`handler`" scripts to create a robust monitoring and alert system. Check scripts can run on many nodes and can report on whether a certain condition is met, such as **Apache** is running. The `handler` scripts can take an action such as sending an alert e-mail.

At its first commit, it was no more than 400LOC, which mostly "ride" on RabbitMQ's routing capabilities—a very smart move!

The real-time processing

It is time to move on to our real-world solution for real-time processing. By now, we should have a great grasp of what Bunny is, and what it allows us to do in the space of RabbitMQ.

To do real-time processing, we will need to be able to use the job queue/workers semantics that we've seen just earlier; but by having this, we may want to just handle the actual logic—what can we do to avoid rewriting the boilerplate that is a RabbitMQ worker?

Furthermore, what can we do to avoid the boilerplate that is a production-grade RabbitMQ worker? Such a production-grade worker should have available configuration, logging, metrics, and a good way to abstract away all of the AMQP/RabbitMQ gritty detail so that we're left with a familiar Ruby-like programming model.

When I started doing this, Ruby didn't have anything to offer. I ended up building my own library, and I've called it `Sneakers`.

Sneakers

Sneakers is defined as a performance background processing library for Ruby and it is exactly that. Sneakers has its place with other scalable solutions on other platforms, such as Python's celery, and Ruby's own `Sidekiq.Sneakers` library uses a hybrid process-thread model, where many processes are spawned and many threads are used per process. Hence, all your cores max out and you have best of both worlds.

Sneakers have processed billions of messages per year at my production projects, handling very critical data, such as billing, logging, and user-tracking data. I've literally put my and my company's money-making data in Sneakers' hands.

> You can always find Sneakers at `http://sneakers.io`.

Installing

Let's see how we can set up our environment for Sneakers worker development. First, install it using Rubygems:

```
$ gem install Sneakers
Fetching: sneakers-1.0.2.gem (100%)
Successfully installed sneakers-1.0.2
Parsing documentation for sneakers-1.0.2
Installing ri documentation for sneakers-1.0.2
Done installing documentation for sneakers after 0 seconds
1 gem installed
```

Running the Sneakers system is composed of specifying the type of work to do `work` is the class of worker and `TitleScraper` is returning root file. `TitleScraper` can include all of our code and its dependencies, and in this case, our actual worker code.

Here's how we run our worker using `sneakers_worker.rb`, which we will write in a few moments (remember to have a local working RabbitMQ broker up):

```
$ sneakers work TitleScraper --require sneakers_worker.rb


          __
      ,--'  >  Sneakers
      `=====


Workers ....: TitleScraper
Log ........: Console
PID ........: sneakers.pid

                    Process control
============================================================
Stop (nicely) ..............: kill -SIGTERM `cat sneakers.pid`
Stop (immediate) ...........: kill -SIGQUIT `cat sneakers.pid`
Restart (nicely) ...........: kill -SIGUSR1 `cat sneakers.pid`
Restart (immediate) ........: kill -SIGHUP `cat sneakers.pid`
Reconfigure ................: kill -SIGUSR2 `cat sneakers.pid`
Scale workers ..............: reconfigure, then restart
```

```
=================================================================

2015-03-07T16:46:47Z p-71531 t-ov268zh7k WARN: Loading runner
configuration...
2015-03-07T16:46:47Z p-71531 t-ov268zh7k INFO: New configuration:
#<Sneakers::Configuration...>
2015-03-07T16:46:47Z p-71531 t-ov268zh7k WARN: Loading runner
configuration...
2015-03-07T16:46:47Z p-71531 t-ov268zh7k INFO: New configuration:
#<Sneakers::Configuration...>
2015-03-07T16:46:47Z p-71531 t-ov268zh7k WARN: Loading runner
configuration...
2015-03-07T16:46:47Z p-71531 t-ov268zh7k INFO: New configuration:
#<Sneakers::Configuration...>2015-03-07T16:46:47Z p-71546 t-ov268zh7k
INFO: Heartbeat interval used (in seconds): 2
2015-03-07T16:46:47Z p-71547 t-ov268zh7k INFO: Heartbeat interval
used (in seconds): 2
2015-03-07T16:46:47Z p-71548 t-ov268zh7k INFO: Heartbeat interval
used (in seconds): 2
```

This is a whole lot of logging. We can see special unique IDs assigned per process and per thread, a logging level, and some internal AMQP detail; these things are expected from a production grade worker class systems.

Next, let's see what it takes to make this quick worker:

```
require 'sneakers'
require 'logger'
Sneakers.logger.level = Logger::INFO

classTitleScraper
include Sneakers::Worker
from_queue 'sneakers_jobs'

def work(msg)
puts "hello sneakers: #{msg}"
ack!
end
end
```

As you can see, we require Sneakers, then a logger in order to configure a good level of logging for the book (we don't want too much logging here!).

From then on, it looks like a plain Ruby. The DSL is specially designed to remind you of other and older background job libraries, such as `delayed_job`, **Resque**, or the now-very-popular (but Redis based) **Sidekiq**.

You make a class with a simple work method; it is your worker, which you can run, test, and maintain in a very streamlined fashion.

Then, tell the class from where to pull messages with `from_queue` and make it into a worker by mixing it with the `Sneakers::Worker mixin` argument. Since we're in full control, we also pass an explicit `ack!`; here, although specifically for your Sneakers workers, ack is implicit.

And for the producer, do as follows:

```
require 'sneakers'
Sneakers.configure
publisher = Sneakers::Publisher.new
publisher.publish('hello', :to_queue => 'sneakers_jobs')
```

Sneakers includes a convenience class called `Sneakers::Publisher`, which you can use explicitly or, if you use **Rails**, you can use the community contributed class called the **Sneakers ActiveJob** interface (this is out of the scope of the book, but feel free to explore it).

In our publisher, we simply create a new one and push a message called 'hello' with a simple target queue instruction called `to_queue`. That's about it.

If we run everything, we get the following:

```
$ruby sneakers_publisher.rb
2015-03-07T17:01:49Z p-74143 t-ow79rce6o DEBUG: Sent protocol preamble
2015-03-07T17:01:49Z p-74143 t-ow79rce6o DEBUG: Sent connection.start-ok
2015-03-07T17:01:49Z p-74143 t-ow79rce6o DEBUG: Heartbeat interval negotiation: client = 2, server = 580, result = 2
2015-03-07T17:01:49Z p-74143 t-ow79rce6o INFO: Heartbeat interval used (in seconds): 2
```

```
2015-03-07T17:01:49Z p-74143 t-ow79rce6o DEBUG: Sent connection.tune-ok with heartbeat interval = 2, frame_max = 131072, channel_max = 65535
2015-03-07T17:01:49Z p-74143 t-ow79rce6o DEBUG: Sent connection.open with vhost = /
2015-03-07T17:01:49Z p-74143 t-ow79rce6o DEBUG: Initializing heartbeat sender...
2015-03-07T17:01:49Z p-74143 t-ow79w6rfs DEBUG: Session#handle_frame on 1: #<AMQ::Protocol::Channel::OpenOk:0x007fbf7383c7c0 @channel_id="">
2015-03-07T17:01:49Z p-74143 t-ow79w6rfs DEBUG: Session#handle_frame on 1: #<AMQ::Protocol::Exchange::DeclareOk:0x007fbf7382f0e8>
2015-03-07T17:01:49Z p-74143 t-ow79w6rfs DEBUG: Channel#handle_frame on channel 1: #<AMQ::Protocol::Exchange::DeclareOk:0x007fbf7382f0e8>
2015-03-07T17:01:49Z p-74143 t-ow79rce6o INFO: publishing <hello> to [sneakers_jobs]
```

This was on the publishing side. Now, let's examine our worker side's log that has been waiting for messages this whole time:

```
2015-03-07T17:01:44Z p-74133 t-ow3bhml78 INFO: Heartbeat interval used (in seconds): 2
2015-03-07T17:01:45Z p-74134 t-ow3bhml78 INFO: Heartbeat interval used (in seconds): 2
hello sneakers: hello
```

So, we conclude our sanity check of making background jobs with Sneakers. There are no connection set up ceremonies; we're left at the same model that we already know from the Ruby world—classes and simple code maintenance.

Sneakers was designed for a polyglot environment in a microservice architecture, where you potentially have many components, platforms, and programming languages and you glue them all together using a unified messaging solution such as RabbitMQ. This is why you should also probably keep your messages easily digestible with a common format such as JSON.

Lambda architecture

As we have seen briefly, Lambda architecture is a useful framework to think about when designing big data applications. Let's take a look at the following diagram:

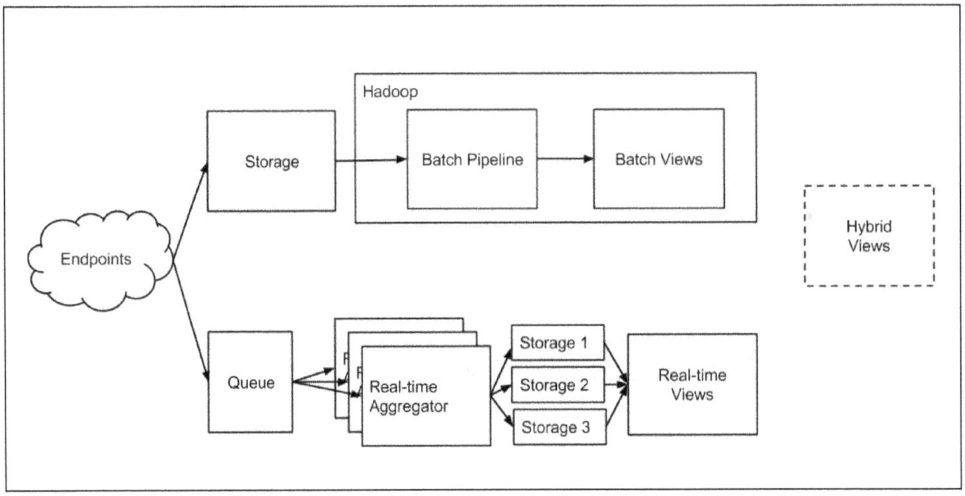

This is the Lambda architecture. It is a bit complex, but rightly so, as it stands as an almost-holy-grail of data processing.

Let's break it down:

- **Endpoints**: This is the web/mobile/any collectors that accepts data from clients. They're in charge to persist these data to a cold storage and a streaming data queue. Essentially, they're making a data split.
- **The real-time processing**:
    - **Queue**: This is the main facility uses for stream-splitting
    - **Real-time Aggregator**: This is the entity that is responsible to feed off each stream and create its own world of data in a custom-specific way
    - **Storage engines**: For each data to be aggregated and queried in a scalable manner, we need to persist custom data to custom data bases
    - **Real-time Views**: This case is same as the custom stores; we probably will need a custom way to query these

- **Batch processing**:
  - **Storage**: This is used for raw data/event storage
  - **Batch Pipeline**: Batch data processing—ETL, aggregations, and view materialization
  - **Batch Views**: Query engine and view layer

The real-time subsystem of Lambda architecture is a secret. We're basically saying that we want to treat a single stream of data *S* in *N* different ways. Maybe, we want to aggregate a sum to count instances and signal alerts and so on. This means we will also need to split the data into *N* similar steams and feed it to *N* different workers.

Know a way to do this? Yes. RabbitMQ will support a fanout, which will duplicate our streams, and it will also support different queues and internal fanout between exchanges. The world of messaging is wide open for us.

The real-time processors

In this chapter, we will focus on a part of Lambda architecture, the meat of it— real-time processors. As you might have guessed, the queuing solution that we will pick is RabbitMQ; now, all that is left is for us to pick a technology or a library to implement the processors in. For this, we'll pick Sneakers.

First, let's sketch out our actual data; let's build a message modeling user-interaction/clicks with the help of the following:

- IP
- User agent
- Channel
- Event type
- Event start
- Event duration
- Event content sample
- Content URL

This is the raw data. Our secret sauce (and probably others') is to take this simplistic data and extract even more from it. Can you see how? Take a look at the following:

- **IP**: Using IP-to-location, we can get information about the location of the user.
- **User agent**: Using device detection, you can learn about the user's devices and the segmentation of the user. For example, if the user using an iPhone 6, he probably is a more of an early-adopter. On some industries (advertising), an iPhone user is worth a lot more than Android.
- **Event content sample**: Although, the user didn't tell us anything about his locale, we can do natural language processing on the item that he was actually looking at, and then we can deduce his language. This is also worth a lot in some industries, and we never need to "bother" the user with useless questions.

So, let's make a sample message and format our messages in JSON:

```
{
  "ip": "8.8.8.8",
  "user_agent": "Mozilla/5.0....ari/534.30",
  "channel": "mobile",
  "event_type": "view",
  "event_start": 1425819361,
  "event_duration": 53,
  "content_sample": "Cistern had graphs back … he past.",
  "content_url": "http://misfra.me/state-of-the-state-part-iii"
}
```

Key performance indicators (KPIs)

KPIs is used to enhance and measure the organization's strategy; so, they must be chosen with accuracy and be set up clearly in order to make them useful. We generally refer to this term as things we would like to measure. In the real-time model, we're inverting the query. We define beforehand the conclusions and things that we would like to deduce from the data. For example, just by seeing this single message, I can already say I would like to understand:

- Average event duration and how long a user spends on an item?
- What percentage of the users use which channel and what users go on Mobile, on Web, and so on?

- Where are most of my users located at?
- When do most users perform actions?
- What is the distribution of content sources my users consume?

Let's take one of these KPIs—averages. What would we need to understand averages? Well, from our basic school education, we know that we need the total number of users and a list of all of their session times. Then, we divide the sum of all session times by the number of users to gain the average session time of all users.

However, when doing real-time computation, an engineer should really understand that since she is dealing with streams—all algorithms should also be adapted to streams. The simple math that we just described expects the data to be ready before calculation.

In the case of Lambda architecture and the real-time processing model, we have a stream of data that we don't know the size of. We don't know when it will end, and most certainly, we can't expect which data we will find on its receiving end.

So, this is where I introduce **moving-averages**. Computing averages seems to be great for a streaming context. You can take a partial average and given a new sample of data, include it into computation, based on the previously consumed data.

Since averages also get "old", we want to drop old samples.

$$SMA_{\text{today}} = SMA_{\text{yesterday}} - \frac{p_{M-n}}{n} + \frac{p_M}{n}$$

That is, if you have *N* samples and a *kth* sample comes along, you take the *AVG(N)* that you already computed, multiply it by the total *N* that you already counted, and then add it to the regular Average formula with your new data sample, but remove the old one.

Building averaging workers

Before diving into our implementation, I'd like to give a little bit insight about Redis. Redis is in-memory data structure store used as database, cache, and message broker. It supports data structures, such as strings, hashes, lists, sets, sorted sets with range queries, bitmaps, hyperlog logs, and geospatial indexes with radius queries.

For now, let's assume that our window is big enough so that we don't need to drop off old samples. Let's build a worker that will sum up our events. First, let's look at the code of such a completed worker:

```
require 'sneakers'
require 'redis'
require 'logger'
require 'json'
Sneakers.logger.level = Logger::INFO

$redis = Redis.new

classAveragesWorker
include Sneakers::Worker
from_queue 'averages_stream'

def work(msg)
event = JSON.parse(msg)
duration = event['event_duration'].to_i
event_count = ($redis.get('event:count') || 0).to_i
event_avg = ($redis.get('event:avg') || 0).to_f

new_count = event_count + 1
new_avg = (event_avg*event_count + duration) / new_count*1.0

    # we are not using #inc, in order to avoid
    # introducing a data race condition.
    $redis.set('event:count', new_count)
    $redis.set('event:avg', new_avg)

puts "Computed average: #{new_avg}, count: #{new_count}"
ack!
end
end
```

So, we build our worker as usual, requiring sneaks and creating a class that mixes in `Sneakers::Worker` argument. We also specify our queue as `averages_worker` field, having assumed that our averaging workers will be facing a dedicated data stream.

Next up, we implement our work method. We parse our JSON message that is pushed onto the queue and extract the `event_duration` field. This field is in second resolution.

Next up, we fetch the existing data from Redis. Redis is a simple key/value store that you can use, which will have very little requirements on your base system. Let's install it now:

Windows

Download the Redis package from `http://redis.io`.

Linux

You can run the following command in order to install Redis on Linux:

```
$ sudo apt-get install redis
```

Mac OS X

With the most useful homebrew, run the following:

```
$ brew install redis
```

You can then run Redis locally as instructed by your own package manager. For more information, use the documentation on `http://redis.io`.

Continuing with the code, we then get and set values onto Redis:

```
event_count = ($redis.get('event:count') || 0).to_i
event_avg = ($redis.get('event:avg') || 0).to_f
```

This means we fetch values from the `event:count` and `event:avg` keys, and if they're nil, we normalize to '`0`', which is acceptable in case of averages.

Next up, we compute the new average based on the old average and store it back to Redis.

If you already know Redis, you also know that there's a command called **INC**. We do not use it because the state that we are modifying must be atomic. We must compute the average and counts atomically, and this is why each worker will explicitly set both overwriting, whatever value were there before on Redis.

We finally print the result for our convenience and end the work with `ack!`.

We also modified our publisher and created `event_publisher.rb` in the following way:

```
require 'sneakers'
Sneakers.configure
publisher = Sneakers::Publisher.new
publisher.publish(File.read('sample_message.json'), :to_queue =>
'averages_stream')
```

Let's see whether it is actually working; start the worker as follows:

```
$ sneakers work AveragesWorker --require averages_worker.rb
```

And now, push a couple messages, as follows:

```
$ rubyevent_publisher.rb
2015-03-08T13:29:13Z p-97040 t-ovcjatdy8 INFO: publishing <{
  "ip": "8.8.8.8",
  "user_agent": "Mozilla/5.0 (Linux; U; Android 4.0.3; ko-kr; LG-
  L160L Build/IML74K) AppleWebkit/534.30 (KHTML, like Gecko)
  Version/4.0 Mobile Safari/534.30",
  "channel": "mobile",
  "event_type": "view",
  "event_start": 1425819361,
  "event_duration": 53,
  "content_sample": "Cistern had graphs back in October 2014. I think
  I used my metricstore package. I'm not sure because I think I was
  switching storage engines every other week! I had both BoltDB and
  SQLite in the source code at some points in the past.",
  "content_url": "http://misfra.me/state-of-the-state-part-iii"
}
>to [averages_stream]
```

If you jump over to the worker terminal, you will see that it computes, saves state, and that everything simply works.

```
2015-03-08T13:29:10Z p-97009 t-ouxjwqw9k INFO: Heartbeat interval
used (in seconds): 2
2015-03-08T13:29:10Z p-97010 t-ouxjwqw9k INFO: Heartbeat interval
used (in seconds): 2
Computed average: 53.0, count: 2
Computed average: 53.0, count: 3
```

That's it. You've build the first piece of the real-time component of your Lambda architecture. What's the best part? It is completely independent. It is loosely coupled, doesn't require anything else—any existing codebase, and it will maintain and evolve separately even by a separate team.

As a bonus, this kind of solution scales to thousands of messages per second using a standard AWS EC2-large server. Once this is deployed, you will probably never need to tend to change it again, it will just keep working.

Building the IP2Location worker

Since everything here is modular, we can continue to build a new worker that does something completely different, but with the same message!

We want to turn every IP into an address by a process called **ip-to-location**.

First, let's make sure we are set up to do ip-to-location. We will use the popular Maxmindip-to-location data base provider. Maxmind sells and also gives a free version of IP databases, which stores a map of all IPs on the Internet to their originating address.

The gem to interact with this database in the Ruby world is called **geoip**. Let's install it:

```
$ gem install geoip
```

Next, you will need a copy of the maxmind database that is available for free at `http://geolite.maxmind.com/download/geoip/database/GeoLiteCity.dat.gz`.

After you have downloaded and extracted it to the same folder as where your code lives in, it is ready to use.

I will start by showing you the completed worker:

```
require 'sneakers'
require 'logger'
require 'json'
require 'geoip'

Sneakers.logger.level = Logger::INFO

$geo = GeoIP.new('GeoLiteCity.dat')

class Ip2locationWorker
include Sneakers::Worker
```

```
from_queue 'ips_stream'

def work(msg)
event = JSON.parse(msg)
target_ip = event['ip']
city = $geo.city(target_ip)
puts city
ack!
end
end
```

Again, simplicity is a key here. We store the geo database within a globally-available `$geo` variable, and then we just extract our IP and query against the database. Also, notice that we gave this worker a stream of its own as a separate queue.

Let's see how this worker performs when a message is pushed:

```
2015-03-08T16:56:41Z p-12153 t-oux39f2z4 INFO: Heartbeat interval
used (in seconds): 2
```

```
#<structGeoIP::City request="8.8.8.8", ip="8.8.8.8",
country_code2="US", country_code3="USA", country_name="United
States", continent_code="NA", region_name="CA", city_name="Mountain
View", postal_code="94040", latitude=37.385999999999996, longitude=-
122.0838, dma_code=807, area_code=650,
timezone="America/Los_Angeles", real_region_name="California">
```

Bam! Within 5 minutes of a bit of glue code and work, we have a production grade worker that can do ip-to-location on demand. Again, no other worker had to be modified; this is pure modular design.

Exploring sneakers

You just saw how to create production grade workers to form a production grade real-time processing pipeline. The idea of production grade shouldn't be taken lightly, because you would spend 20% of your time on your logic and 80% of your time making operational decisions and steps, such as deploying, figuring out why stuff doesn't work, and so on. Like the old saying—the devil is in the detail, and the goal of Sneakers was to package up solutions for most of these details out of box.

Next up, we will touch on a few more of Sneakers' production-oriented features.

Timeouts

You can't have workers take jobs that hold them up forever. You must bind the resources you are using. In addition to holding up RAM and CPU, one less-obvious resource is time, and as they say—time is money.

A Sneakers worker has configuration to limit this with the `timeout_jobs_after` flag, where you specify the number of seconds you allow this job to take at the maximum.

For example, take a look at the following code:

```
classProfilingWorker
include Sneakers::Worker
from_queue 'downloads',
:ack => true,
:timeout_job_after => 1
```

And you can also configure this globally, like this:

```
Sneakers.configure :timeout_job_after => 1
```

Job handling

Each job or task can be handled differently in your `work` method. You can signal back to RabbitMQ, which is the result that you arrived at as far as messaging is concerned:

- The job is done—`ack!`
- The job has failed (by design)—`reject!`
- The job has failed (an exceptional error)—`nack!`

Metrics

Even without you knowing, Sneakers will automatically measure critical information about your background jobs, such as message sizes and average, time to complete, and so on. Sneakers exposes it to a default metrics handler which is a no-op.

However, Sneakers comes with three metrics handlers out of box:

- **Logging metrics**: This spits metrics to your log console
- **Newrelic metrics**: This pushes metrics to New Relic, a popular metrics company

- **Statsd metrics**: This pushes metrics to your Statsd server (Etsy's open-source metrics aggregation daemon)

You can plugin the one that you wish to use by specifying it in the central configuration:

```
Sneakers.configure(:metrics =>
Sneakers::Metrics::LoggingMetrics.new)
```

This will cause all of the metrics to be spilled out to your terminal, which is useful when you're debugging or developing locally on your machine.

Summary

Using RabbitMQ from Ruby is a pleasure. I can personally say that this combination represents a set of two technologies that look so fitting together that one may think they belong together.

We saw how to develop easy workers and simple messaging topologies using the AMQP semantics, Bunny, and a Ruby library for RabbitMQ, which is one of the most advanced out there over multiple platforms and programming languages.

We also saw how to with the little details of AMQP, assuming you want to build on the worker abstraction with background jobs, and you want to build a production proven worker solution with Sneakers.

We also explored Lambda architecture, and we saw how Sneakers helps you build more content into it and how it is able to invest in it for your data processing products.

I can say assuredly that you can take these solutions today and build your data science pipelines on them. It's what I have been doing for the last few years myself. The next chapter will cover Python client programming for RabbitMQ.

11
Python Client Programming

In this chapter, we will continue with our exploration into RabbitMQ clients, but this time in Python.

Python provides an almost perfect environment into this work, because it is dynamic in nature and over the (many) years, it has established itself as an almost de-facto toolset for backend engineers and data scientists. It is only fitting that during earlier time, it has allowed great tooling to evolve, and we'll explore them here.

In this chapter, you will learn the following topics:

- Using RabbitMQ from Python with the help of a library called Pika
- Using Pika to implement a sample use case project
- Exploring how Pika will help you tackle almost any AMQP-related task
- Understanding Celery—a powerful background task library
- Understanding how the work we did so far adapts into Celery, why it is better, and how to use its more-advanced usage

Case study

For this study, we'll choose the world of web scraping. Python is universally known for this use, because if you're a python programmer, you're one of the following:

- A full-stack Python programmer, developing client and server applications
- A data scientist, using NumPy, scikit-learn, and so on
- A systems programmer, as Python comes standard with most, if not all, Linux distributions
- A big-data programmer, utilizing Hadoop and Python-Hadoop streaming or Python for building map-reduce jobs

A successful web scraping infrastructure for a successful product generates a lot of data (a big data programmer) that we can learn from (a data scientist); it is rather taxying to properly build at scale (a systems programmer), and somehow, you'd want to turn all this into a product (a full stack Python programmer).

As you see, once you cover web scraping, you can build a lot of products; you can also get into a lot of technological domains in the same programming language and platform. And if this is not enough, any of these domains are a hot recruitment targets today.

I hope this is enough motivation to explore Python and how RabbitMQ fits into this world.

Getting Python dependencies

One of the easy things do in Python is getting dependencies—the idea of using the libraries that you've chosen to integrate into your code, their dependencies, and well, you get the picture.

Python comes with a few tools to do this, and you can pick what you wish based on what works for you the best. Some of these are as follows:

- `pip`: This is a Python dependency manager
- `easy_install`: This is yet another Python dependency manager
- `apt-get`: This is available on Debian and Ubuntu Linux; the standard package manager can also install Python packages

Pika

We'll start with a Python RabbitMQ client called **Pika** that offers a good balance between low-level and high-level APIs and developer happiness. Other libraries such as `py-amqplib` and `txAMQP` are also good to use, but they provide a different kind of balance (mostly towards low-level) and will not be covered here. As always, you are free to try them out, but my recommendation is that you try to do it only after you have grasped a good understanding of Pika.

Installing Pika

Let's install Pika on our workstations.

Before this, let's ensure we have an appropriate package manager already provisioned on our machines.

On Linux (Ubuntu), type this at your terminal:

```
$ sudo apt-get install python-pip git-core
```

On Windows, type this:

First, make sure to use the **setup tools** installer at `https://pypi.python.org/pypi/setuptools`. Then, run this command:

```
>easy_install pip
```

Now, we can install Pika.

Type these commands into your machine's terminal:

On Linux (Ubuntu) type this at your terminal:

```
$ sudo pip install pika
```

On Windows type this:

```
>pip install pika
```

Our first Pika client

Let's verify that we've got Pika installed properly by building a simple client and server, or in other words, a *producer* and *consumer* that connect to our local RabbitMQ broker instance. Here's how our `producer.py` file looks like:

```
#!/usr/bin/env python
importpika

print "* Connecting to RabbitMQ broker"

connection = pika.BlockingConnection(pika.ConnectionParameters(host='
localhost'))

channel = connection.channel()

channel.queue_declare(queue='pages')

#default empty exchange with routing key equal to the queue name #
will route the message to that queue

channel.basic_publish(exchange='',
routing_key='pages',
```

```
body='testing: 1, 2, 3')

print "* Done sending!"
connection.close()
```

Let's overview what we've accomplished here. Shortly after the Python Shebang (`#!/usr/bin/env python`), we import Pika. This lets us use the RabbitMQ API as described by Pika's API documents.

We start, as always, by opening a connection; in this case, a blocking connection. There's also a non-blocking connection called a **Select Connection** whose name is inspired from kpoll/epoll operating system mechanisms, which allow us to do asynchronous I/O very easily and efficiently. For now, we'll stay with the still-well-performing Blocking Connection.

Next up, we will build a *Channel*, which is what we're going to send data on with AMQP to declare a queue. A note about declaring a queue—we can either let the process use and expect a queue to be there when it gets started, and, if it is missing, it should crash; or, we can instruct the process to build the queue if it is missing.

The way we chose to negotiate this dilemma is through building the queue if it is missing: hence, Declare. It is best practice to do this, and there is no harm in creating a missing queue; however, there is a deep pitfall here to pay respect to. That is, a queue or exchange, once declared dynamically is created with the default properties of the given library (in this case, Pika). And sometimes, for example, in the case of an exchange in order to change a property, you'll have to recreate the object that you're unhappy with.

Let's run this code, and later, continue going over it:

```
$ python producer.py
Traceback (most recent call last):
  File "producer.py", line 5, in <module>
    host='localhost'))
  File "/usr/local/lib/python2.7/site-packages/pika/adapters/blocking_
connection.py", line 130, in __init__
super(BlockingConnection, self).__init__(parameters, None, False)
  File "/usr/local/lib/python2.7/site-packages/pika/adapters/base_
connection.py", line 72, in __init__
on_close_callback)
  File "/usr/local/lib/python2.7/site-packages/pika/connection.py",
line 600, in __init__
self.connect()
  File "/usr/local/lib/python2.7/site-packages/pika/adapters/blocking_
connection.py", line 230, in connect
```

```
error = self._adapter_connect()
  File "/usr/local/lib/python2.7/site-packages/pika/adapters/blocking_
connection.py", line 301, in _adapter_connect
raiseexceptions.AMQPConnectionError(error)
pika.exceptions.AMQPConnectionError: Connection to fe80::1%lo0:5672
failed: [Errno 61] Connection refused
```

Whoa! It crashed on us. By the looks of, after following the stack trace, the most familiar looking keyword here is `AMQPConnectionError`, and surely enough, we see `Connection refused` following after it.

Our RabbitMQ host isn't really running at all. I'll let you amend the problem by running your RabbitMQ host while I fix mine.

This method of failing fast is great for learning. I always like to explore by surprise, and surely, familiarizing yourself with the common errors of a library or a framework very early in the learning process is one of the better programmer's secret weapons.

A consumer

Now that we've verified that we've a RabbitMQ broker well and running on our local machine (on local host), let's continue with the following code:

```
$ python producer.py
* Connecting to RabbitMQ broker
* Done sending!
```

So, this time it works. There should be a message waiting for us in the `pages` queue. Let's quickly build a consumer to see whether we can fetch the items in the queue.

First though, let's make sure we didn't goof up, and there are messages in our newly created queue. We'll use the immensely helpful `rabbitmqctl` tool:

```
$ rabbitmqctllist_queues
Listing queues ...
logs  0
pages  1
test_stress  0
testqueue  0
```

I have emphasized the queue that we are interested in, the `pages` queue, and you can easily see it contains a message.

Now, for the consumer, we've the following:

```
#!/usr/bin/env python
importpika

def handler(ch, method, properties, body):
print "-> Handled: [%s]" % (body)

connection = pika.BlockingConnection(pika.ConnectionParameters(host='
localhost'))

channel = connection.channel()

print '* Handling messages.'

channel.basic_consume(handler, queue='pages', no_ack=True)

channel.start_consuming()
```

Let's go over the code. We'll again start with Shebang and import Pika, but the next interesting bit is our handler, which we'll use to process our messages with. This is where the meat of your consumers will usually be, all the rest will find its space within the boilerplate category.

Unlike the producer, this consumer assumes there already is a queue called `pages` on the broker.

Next, we will introduce the `a basic_consume` bit. This is where Pika and other similar libraries for other languages shine. Notice that we're working without being acknowledged by specifying `no_ack=True`. We'll come back to this later.

Where these kinds of mid-level libraries shine is by allowing you to just specify a handling function which takes a message and processes it, and they'll take care of managing the actual, which is perhaps less interesting to you, and the technical bits of AMQP against the broker.

Introducing the web scraper

Let's review a simple web scraper architecture:

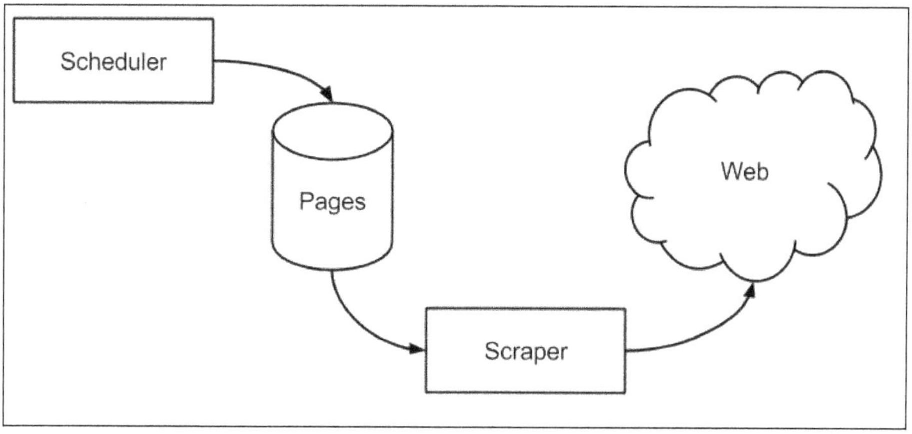

Scheduler

The web changes often. It is a huge and dynamic beast. The **scheduler** is responsible to make sure that the scraper will always represent data that is fresh and not stale. It is free to do so by deciding at what rate to scrape it for each website or the page that is being scraped; in other words, when is the next scraping going to happen.

In reality, you would want the scheduler to feed from a persistent data store that holds all sources and their upcoming scraping time.

For example, you could hold a record that specifies that the website `acme.org` will have to be scraped once every 5 minutes. You could even pour some more sophistication into it. You can state that `acme.org` has to be scraped every 5 minutes at day time, but at night time, in order to save your resources, a 30-minute cycle would be good enough.

Whatever your scheduling policy is, it is encapsulated within the Scheduler domain.

Scraper

A **scraper** is responsible to get a URL, fetch its content over the network, parse it, and possibly extract important information from it.

After extraction, it persists or outputs the findings in any agreed-upon way.

The scraper should be stateless, because in scale, you would want to deploy as many of these as required (think about AWS autoscaling), and perhaps you could kill off some of them, when the processing power isn't really needed anymore.

This is why the worker model is perfect here.

Having said that and being armed with our producer/consumer models, we can now easily step it up a bit and build a web scraper.

Implementing the scheduler

We will start off with the base code of the producer. If you can think about it, you'll find that we're only missing the scheduling piece from it.

For actual scheduling, we'll use the cool `schedule` library. Let's get it with `pip`:

```
$ pip install schedule
Downloading/unpacking schedule
  Downloading schedule-0.3.1.tar.gz
  Running setup.py (path:/private/var/folders/gw/xp4xsqt97957cc7hcgxd
0w0c0000gn/T/pip_build_dotan/schedule/setup.py) egg_info for package
schedule

Installing collected packages: schedule
  Running setup.py install for schedule

Successfully installed schedule
```

And we're done. Here's how we can use `schedule` (from the documentation):

```
import schedule
import time

def job():
print("I'm working...")
```

```
schedule.every(10).minutes.do(job)
schedule.every().hour.do(job)
schedule.every().day.at("10:30").do(job)
schedule.every().monday.do(job)
schedule.every().wednesday.at("13:15").do(job)

while True:
schedule.run_pending()
time.sleep(1)
```

Because `schedule` is so developer-friendly, it is quite easy to understand what's going on; it is almost plain English.

In the previous example, we imported `schedule` and we' defined a function called `job` that we'll use to run. Sounds like a create place to push a message, right?

Continuing, we're sticking `schedule.run_pending()` within a never-ending while loop. 'This is how we are going to get an always-on daemon process.

I think we can now build a proper scheduler:

```
#!/usr/bin/env python
importpika
import schedule
import time


urls = [ "http://ebay.to/1G163Lh" ]

print "* Connecting to RabbitMQ broker"

connection = pika.BlockingConnection(pika.ConnectionParameters(host='
localhost'))

channel = connection.channel()

channel.queue_declare(queue='pages')

def produce():
forurl in urls:
print("* Pushed: [%s]" % (url))
channel.basic_publish(exchange='', routing_key='pages', body=url)


schedule.every(10).seconds.do(produce)
```

```
while True:
schedule.run_pending()
time.sleep(1)

connection.close()
```

This example proves how simple Pika can make our lives. It's simply a glue that holds a schedule and message together.

Let's go over it now.

We'll be holding our intended URLs in a simple URL array. This is the place where you can swap it out with a real persistence layer (for example, either via Postgres or MongoDB).

For the sake of the scope of this book, we'll imagine that this URL array is our persistence layer.

Next up, we'll opening a connection, creating a channel, and finding or creating a queue called `pages`, just as we did before; however, this time we mean business. This time, we need this queue to hold our real URLs so that the next guy in line—Scraper will be able to digest them.

In order for the `schedule` library to work, we need to define a function that will pass to it. This function, as hinted before, is actually going over the URLs and pushes them one by one (in the example, for simplicity, we are holding just one) onto the pages queue.

In the following lines, we are actually defining the schedule and building a small harness that will check the schedule, execute schedule items, and repeat this process forever; this is what the while loop is doing there.

In any case, if ever the while loop stumbles on an exception, we are sure to close a connection so that we let the broker know that it can clean up resources and, well, it contributes for us being a well-behaved RabbitMQ citizen.

Implementing the scraper

Scraper would be a system of copying content of other websites using web scraping. First, we want to state a few of the things that we want to accomplish:

- Downloading a web page
- Parsing HTML
- Cherry-picking attributes from the HTML
- Saving the results

For a modern way to fetch content from the web, we will avoid the standard `urllib` library and go directly with the nicer `requests` library from the Python community.

For parsing and drilling into web pages, we'll use the almost de-facto library for this in the Python world — `BeautifulSoup`.

Let's fetch these via `pip`:

```
$ pip install requests beautifulsoup
Requirement already satisfied (use --upgrade to upgrade): requests in
/Library/Python/2.7/site-packages/requests-2.2.1-py2.7.egg
Downloading/unpacking beautifulsoup
  Downloading BeautifulSoup-3.2.1.tar.gz
  Running setup.py (path:/private/var/folders/gw/xp4xsqt97957cc7hcg
xd0w0c0000gn/T/pip_build_dotan/beautifulsoup/setup.py) egg_info for
package beautifulsoup

Installing collected packages: beautifulsoup
  Running setup.py install for beautifulsoup

Successfully installed beautifulsoup
Cleaning up...
```

And now, let's sketch out the scraper based on our consumer skeleton:

```
#!/usr/bin/env python
importpika
import requests
fromBeautifulSoup import BeautifulSoup

def handler(ch, method, properties, url):
print "-> Starting: [%s]" % (url)
    r = requests.get(url)
soup = BeautifulSoup(r.text)
print "-> Extracted: %s" % (soup.html.head.title)

print "-> Done: [%s]" % (url)

connection = pika.BlockingConnection(pika.ConnectionParameters(host='
localhost'))

channel = connection.channel()

print '* Handling messages.'
```

```
channel.basic_consume(handler, queue='pages', no_ack=True)

channel.start_consuming()
```

We're going to go over the code, but worry not, since every library that we've used is pretty awesome, our code is highly readable and concise.

We start again with our Shebang and importing `Pika`, `requests`, and `beautifulsoup`. Next up, we beef up our handler from the previous consumer skeleton such that, as promised, all the real logic is contained within it.

Fetching a URL is made very easy with `requests`; the URL's content is available on the response object return to use by the `get` call within the `text` field. This is a simple text, non-parsed and non-digested.

We'll use `bautifulSoup` to turn this raw text into an HTML tree so that we can drill into the meaning from code, rather looking at an array of characters.

Accessing the title is easy with beautifulSoup; by specifying `soup.html.head.title`, we get prime access to it, and all that's left to do is output it somewhere.

We are skipping storing the findings (such as the title) to a persistent store for the sake of brevity. As with the scheduler persistent store, a good look at Postgres or MongoDB will make sense here, but we'll skip it for the scope of this chapter and simply output to the standard output.

Running the scraper

Let's run our scheduler first. Wait a bit, and it will start pushing URLs for the scraper to bite at:

```
$ python scheduler.py
* Connecting to RabbitMQ broker
* Pushed: [http://ebay.to/1G163Lh]
* Pushed: [http://ebay.to/1G163Lh]
* Pushed: [http://ebay.to/1G163Lh]
* Pushed: [http://ebay.to/1G163Lh]
```

On the other end, we'll start our scraper. Feel free to start it on a different terminal, and position it such that you'll have parallel visuals of both the scheduler and scraper.

Your scraper will immediately go to work:

```
$ python scraper.py
* Handling messages.
-> Starting: [http://ebay.to/1G163Lh]
-> Extracted: <title> Killer Rabbit of Death w Pointy Teeth Monty
Python Blinking Red Eyes | eBay </title>
-> Done: [http://ebay.to/1G163Lh]
-> Starting: [http://ebay.to/1G163Lh]
-> Extracted: <title> Killer Rabbit of Death w Pointy Teeth Monty
Python Blinking Red Eyes | eBay </title>
-> Done: [http://ebay.to/1G163Lh]
```

We see that it has downloaded a page, and we are actually pulling out the `<title>` element from each page! The way from here to pulling out product details and building a sophisticated data-driven product based on Ebay's data (of course, please adhere to Ebay's terms of service) is very, very short.

Handling failure

When building a robust system such as a scraper that needs to be on 24 x 7 and perform reliably, several things can fail, but surprisingly, they can also be easily fixed:

- Website errors
- Network errors (RabbitMQ connectivity)
- Programmer error (typos)

Handling programmer errors can be easily fixed by having proper testing, so 'I'll leave that one to you.

However, we can divide the web and network errors into two classes:

- Persistent errors
- Transient/temporary errors

A persistent error is something that's not fixable with ease, for example, a disk failure. A transient error is something that'll probably be fixed without our interfering. A website going down and returning errors to our scraper download code is not something that we can fix; however, since it is transient, we can save the drama for later, and retry it on a later occasion.

A network glitch, disconnecting our TCP socket and causing the connection to our RabbitMQ broker to break is also transient, and we can solve it by retrying the connection again.

Using acknowledgement

In order to properly handle the case where a web server that we need to fetch data from goes down, we need to understand one thing. Once a worker fetches a message from RabbitMQ with a `no_ack` model, which we have been using so far (feel free to reread the code if this is the first time you've seen it), and if it fails, it will take the message with it.

Unless it will only take the message with it once it is really finished working with it. For this to work, we need to use the acknowledgement model from RabbitMQ.

Each worker, once *really* done with a message, must acknowledge the message back to RabbitMQ, and *only then* will RabbitMQ sign the message off and remove it from the queue.

Let's update our worker code to acknowledge messages:

```
#!/usr/bin/env python
importpika
import requests
fromBeautifulSoup import BeautifulSoup

def handler(ch, method, properties, url):
print "-> Starting: [%s]" % (url)
    r = requests.get(url)
soup = BeautifulSoup(r.text)
print "-> Extracted: %s" % (soup.html.head.title)

ch.basic_ack(delivery_tag = method.delivery_tag)
print "-> Done: [%s]" % (url)

connection = pika.BlockingConnection(pika.ConnectionParameters(host='
localhost'))

channel = connection.channel()

print '* Handling messages.'

channel.basic_consume(handler, queue='pages', no_ack=False)

channel.start_consuming()
```

I have emphasized the parts that have changed. We instruct RabbitMQ to use acknowledgements with `no_ack=false`, which really is a negative-negative and means `ack=True`.

We're also using the `basic_ack` API from Pika in order to transmit an ACK frame back to RabbitMQ.

This way, if a worker crashes due to any kind of exception, the message remains on the Queue and a new, fresh worker will be able to have a go at it instead.

The Pika API

In this section, we'll cover the various knobs, settings, and API surface area that Pika exposes to you. The programmer Pika is a python implementation of the AMQP 0-9-1 protocol that tries to stay fairly independent of the underlying network support library. Pika doesn't require threads. It takes care of to forbidding them either. The same goes for greenlets, callbacks, continuations, and generators. Pika is available for download via PyPI and can be installed using `easy_install` or `pip`:

```
pip install pika
```

You can also use this:

```
easy_installpika
```

Connecting

There are two ways to set up a connection with Pika. One is to explicitly specify the kind of options you want and expect RabbitMQ to respect as part of a `ConnectionParameters` object, and the other is by specifying a URL that lines out all of the various parameters that you'd like.

Specifying a connection option through a unified URL is a lot more useful these days, as most PaaS platforms, such as Heroku and their add-on partners (such as a RabbitMQaddon), expect you to do it in this way to the promote dynamic behavior of your application by setting a simple environment variable.

Let's start by showing a few examples of the URL parameters configuration option:

```
amqps://www-data:rabbit_pwd@rabbit1/web_messages
```

This URL represents a simple authenticated connection, a host named `rabbit1` and a virtual host called `web_messages`:

```
amqps://www-data:rabbit_pwd@rabbit1/web_messages?heartbeat_interval=30
```

This URL represents a simple authenticated connection, a host named `rabbit1`, a virtual host called `web_messages`, and an explicit `heartbeat_interval` of 30 seconds:

```
amqp://www-data:rabbit_pwd@rabbit1/web_messages?heartbeat_
interval=30&ssl_options=%7B%27keyfile%27%3A+%27%2Fetc%2Fssl%2Fmykey.pe
m%27%2C+%27certfile%27%3A+%27%2Fetc%2Fssl%2Fmycert.pem%27%7D
```

This URL represents a simple authenticated connection, a host named `rabbit1`, a virtual host called web_messages, and an explicit `heartbeat_interval` of 30 seconds; an explicit SSL certificates for the secure connection setup.

To use the URL parameters method, we simply use Pika:

```
pika.URLParameters('amqps://www:pwd@rabbit1/web_messages')
```

Here are some other options you can set up at the query-param level:

- `backpressure_detection`: This disabled by default. Pass a value of that specifies how to handle clients that are too fast (previously, flow control in RabbitMQ).
- `channel_ma`: This is the maximum number of channels allowed for this connection.
- `connection_attempts`: This is default `1`.
- `frame_max`: This is maximum frame size and is useful for performance tuning.
- `heartbeat_interval`: This is the client/server heartbeat interval. In the past, a small value of 5 seconds used to be good, but today, it is encouraged to go with a higher value of `30`.
- `locale`: This is the client locale and is useful if you use a different locale.
- `retry_delay`: This is the time of seconds to wait between connection retries. It usually goes with connection_attempts.
- `socket_timeou`: default `0.25`.
- `ssl_options`: URL encoded dict of the following keys: `ca_certs`, `cert_reqs`, `certfile`, `keyfile`, `ssl_version`.

The next way to connect to a RabbitMQbroker is via an explicit `ConnectionParameters` object; which is much like what we did so far in our scraper project.

We initialize a ConnectionParameters object simply with Pika:

```
pika.ConnectionParameters(host='localhost')
```

However, once we customize it, we can use the same parameter that we just described for the URL parameters method:

```
pika.ConnectionParameters(host='localhost', heartbeat_interval=30,
retry_delay=2)
```

In either case, once you have a parameter's object in your hands, you can pass it down to the actual connection strategy that you've chosen, and then grab a connection:

```
connection = pika.BlockingConnection(pika.ConnectionParameters(host='
localhost'))
```

Using connection adapters

In the previous examples, BlockingConnection represents a connection adapter. A connection adapter is the abstraction that Pika uses in order to hide away the actual strategy it uses for connections. That is, whether it is being a blocking connection, an async I/O event loop, or plugging into the back end that you're using, such as Tornado or Twisted.

To cover most cases, you should only focus on the standard BlockingConnection, and the SelectConnection adapters.

BlockingConnection

Let's survey the API of BlockingConnection (some of these may directly be relevant to other types of connections, such as SelectConnection, since they are both a connection). Following are the parameters of BlockingConnection:

- add_backpressure_callback: This adds a callback to be called when the client experiences a backpressure event from the broker
- basic_nack: This finds out if the broker supports nack with this
- channel: This creates a new channel
- close: This disconnects and returns reply_code and reply_text (both are optional)
- is_closed/is_closing/is_open: This finds out whether the connection is open, closed, or is closing with these

BlockingChannel

Let's survey the API of `BlockingChannel`. This channel is the concept that you'll work against mostly in your programming with RabbitMQ:

- `add_on_close_callback`: This adds a callback to be called when the channel gets closed.
- `add_on_flow_callback`: This adds a callback to be called when the client receives a flow control event.
- `add_on_return_callback`: This adds a callback to be called when the publishing client gets a rejected publish from the server. This is a useful callback to set up, and many don't really treat failure for publishing with RabbitMQ. It can create a very creepy environment for a really hard bug in production.
- `basic_ack / basic_nack`: This one is important. When you're using acknowledge in your message processing semantics, you'll have to use this. When calling this, you should always specify `delivery_tag`, which you'll get on the consuming callback handler.
- `basic_consume`: This is another big one parameter. When developing consumers and workers, it is most likely that you'll end up just using `basic_consume`. The API is tight enough to suffice in most use cases. With `basic_consume` you should also list out the following:
    - The `consumer_callback` that is your handler function
    - The `queue` that is your queue name
    - The `no_ack` that tells the broker whether we want to acknowledge or not (Boolean)
    - The `exclusive` means 'not to allow any other on this queue (boolean)
    - The `consumer_tag` is your own consumer tag, mostly don't use this
    - The `arguments` is custom arguments for consume, mostly don't use this
- `basic_get`: This is just as `basic_consume`, but it gets a single message right then and there, including the `queue` and `no_ack` parameters.
- `basic_publish`: This is the main entry point to push messages onto the RabbitMQ broker. Here, we have several important parameters to describe; they're as follows:
    - `exchange`: This is used to publish.
    - `routing_key`: This is the routing key to bind on.

- `body`: This is the message body.
- `mandatory`: If this is false, the server silently drops a message that cannot be routed to a queue; otherwise it will signal the client.
- `Immediate`: Same as preceding, but now the server will queue the message with no guarantees.
- `basic_qos`- Through this API, the client can control the flow of messages in order to tweak and control the overall performance of processing messages. It can tell the broker to send less messages on a batch, a smaller or bigger messages batch size, and can tell whether to apply to all channels:
- `prefetch_size`: This is the window size in terms of message size. It is invalid when specifying `no_ack`.
- `prefetch_count`: This is the window size in terms of all the messages. It is invalid when specifying `no_ack`.
- `all_channels`: This applies rules to all channels.
- `basic_recover`: This asks the broker to redeliver all unacknowledged messsages.

- `basic_reject`: This is used to reject a message against the broker. Must supply a `delivery_tag` you can also tell the broker to requeue with the `requeue boolean` flag.

Declaring queues and exchanges

In the next few bits, we'll look at the Pika API designated for creating or declaring queues and exchanges. You'll usually make these kinds of calls at the prolog of your consumers or producers, and they will mostly feel like a 'setup code'.

Let's take a look at these now:

- `exhange_declare`: This creates an exchange if it doesn't already exist. Note that if one exist and you are specifying a new one with different parameters, there will be an error representing that. Here are the important parameters that you can specify:
  - `exchange`: This is the exchange name
  - `exchange_type`: This is the type to use (direct, and more); consult the more-detailed RabbitMQ docs for the types and their semantics
  - `passive`: This is used to see if an exchange exists, but it doesn't create one

    - `durable`: This is used to survive a broker reboot—persistent exchange
    - `auto_delete`: Remove this when done using it (no queues bound)
    - `internal`: This can only be published into by other exchanges

- `queue_declare`: This creates a queue if it is not existing with a specified sharing, durability, and other properties, shown as follows:
    - `queue`: This queues names
    - `passive`: This is passive name
    - `durable`: This survives reboots—persistent
    - `exclusive`: This allows you to share between consumers
    - `auto_delete`: Delete this after a consumer disconnects
- `queue_bind`: This binds a queue to a specified exchange. You should provide the following parameters:
    - `queue`: This is the queue name
    - `exchange`: This is the exchange name
    - `routing_key`: This is the routing key
    - `no_wait`: Do not wait for a bind ok

This finalizes most of the API that you will use on a day-to-day basis. There's really not much to it, since Pika has been carefully crafted to simplify AMQP, as did other Pika clones (or maybe Pika is a clone of?), such as Ruby's immensely popular `bunny` library, which I'm deeply fond of.

There are other edge-case API endpoints for the Pika Channel and Connection abstractions, such as deletion of queues, exchanges, unbinding, and cleanups, and more. Should you wish to explore further, refer to the official Pika API documentation.

Authentication

When adopting RabbitMQ universally over your architecture and especially in the enterprise environment, security and authentication comes up immediately.

A few organizations would never even adopt a technology unless it has some widely recognized tiers of security built into them. This is why RabbitMQ grew to support connection authentication and SSL across the board.

Plain credentials

As we've seen with the URL Parameters before, we can use HTTP Basic authentication lined right there on the URL that we pass to Pika. However, we can also use a more programmatic approach using the `PlainCredentials` object:

```
importpika

credentials = pika.PlainCredentials('www', 'pwd')
parameters = pika.ConnectionParameters('rabbit1',
5672
'/',
credentials)
```

From here now, we can pass `parameters` back to our connection.

SSL and external credentials

For those requiring a more secure authentication model, RabbitMQ and Pika allow an SSL-based connection.

Since we haven't seen it before, let's describe how we can get a secure connection through the URL Parameters method. To do this, we need to use a special `amqps` scheme (the part that specifies what we know as the protocol in the URL):

```
amqps_URI       = "amqps://" amqp_authority [ "/" vhost ]
```

This means we can draw up URLs like this:

```
amqps://user:pass@host:10000/vhost
```

Certificate authentication

To get at this kind of functionality programmatically, we need to use the External Credential object. We'll also take advantage of this example to show how to specify custom SSL certificates along the line:

```
ssl_options = ({"ca_certs": "caroot.pem",
                "certfile": "client.pem",
                "keyfile": "key.pem"})

parameters = pika.ConnectionParameters(
host,
      5671,credentials=ExternalCredentials(),
ssl=True,
ssl_options=ssl_options)
```

Here, we are specifying `ssl_options` including CA-Root, a Client pem, and our actual key. We are switching credentials to `External` with `ExternalCredentials` since we have the help of PKI (private keys and certificates), which is arguably much stronger than a simple user/password combination.

Note that if we double-back to the URL method, we can again opt to use the pem keys and certificates to specify our SSL options in the same way. To do this, we'll need to encode our `ssl_optionsdict` right there on our URL using `urllib`:

```
url = urllib.urlencode({'ssl_options': {'ca_certs': 'caroot.pem',
'certfile': 'client.pem', 'keyfile': 'key.pem'}})
```

Background processing

This concludes our exploration of the Pika API. Still, there are other Pika gems to be found within its API, but rest assured, not many. You're free to explore the nitty-gritty details at the official ReadTheDocs API, available at `https://pika.readthedocs.org`.

We'll continue by jumping *a lot higher* in the abstraction model and move up to the big shot tools that'll allow us to do background or queue-based processing without really getting our hands dirty with the AMQP protocol or even close.

Having just explored the Pika API gives you immense advantage over anyone else on this same space, since you know how stuff works, and well, sometimes, you don't need a very big hammer for a very small nail.

And now, let's explore big hammers.

Celery

Celery is a Distributed Task Queue. What this means is that in the context of RabbitMQ and AMQP, it takes the entire AMQP model and shapes it; it folds only the best ideas from it into providing a world-class, production-grade background queue library for you to use.

Celery allows for swappable backends, and one of them is RabbitMQ, which we'll explore here.

Celery is mind-numbingly used almost everywhere in nearly every Python-based company or start-up to do background jobs; it's also used in big corporations, such as Mozilla; see more information at: `https://github.com/celery/celery/wiki#companieswebsites-using-celery`.

Hopefully, you're excited about exploring Celery as I am; I hope you'll be amazed at Celery's conciseness.

Installation

First, let's install `celery` as 'we did before with pip:

```
$ pip install celery
```

Then, let's make sure we have a working installation by verifying with a simple client:

```
from celery import Celery

app = Celery('pages_celery', broker='amqp://guest@localhost//')

@app.task
def work(msg):
printmsg
```

Type or paste this code into a file named `celery_client.py`. Now, let's run it:

```
$ python celery_client.py
```

If you get no errors you're good to go; however, notice that nothing happens. Celery is a bit different because you need the Celery runner in order to actually make it work against a broker; it's different from what we've experienced with Pika, where you just ran your code and it connected, fetched, and processed messages right off the bat.

Let's run the client properly in the same `celery_client.py` folder; run this:

```
$ celery -A celery_client worker -loglevel=info
```

This means that celery starts at worker mode using our `celery_client` module, and it uses an `info` log level to work with. You will then see this nifty Celery banner:

```
-------------- celery@jondot-mbp.local v3.1.17 (Cipater)
---- **** -----
--- * ***  * -- Darwin-13.4.0-x86_64-i386-64bit
-- * - **** ---
- ** ---------- [config]
- ** ---------- .> app:          tasks:0x104c8abd0
- ** ---------- .> transport:    amqp://guest:**@localhost:5672//
- ** ---------- .> results:      disabled
- *** --- * --- .> concurrency: 4 (prefork)
-- ******* ----
--- ***** ----- [queues]
-------------- .> celery           exchange=celery(direct) key=celery
```

Given that the following log information is not packed with any error ones, you're good to go. However, you may or may not notice that there's no work really happening just yet. Let's see how to throw jobs at this worker.

We'll use the interactive Python shell to push an item:

```
$ python
Python 2.7.9 (default, Dec 13 2014, 22:35:32)
[GCC 4.2.1 Compatible Apple LLVM 6.0 (clang-600.0.56)] on darwin
Type "help", "copyright", "credits" or "license" for more information.
>>>fromcelery_client import work
>>>work.delay("hello world")
<AsyncResult: 85eb0e88-cc6a-4c0a-a7e5-dcf8fabd810a>
```

We see here that we can import the same bit of code that we just created for the worker, and use it from the producer side. This creates a reality where the same code is maintained both for worker and work invocation; and overall, it makes a cleaner code experience for a given team.

On the other end of the queue, let's switch over to the terminal where our worker is still running and verify that a task has been executed:

```
[2015-03-01 11:51:55,816: INFO/MainProcess] Received task: celery_
client.work[85eb0e88-cc6a-4c0a-a7e5-dcf8fabd810a]
[2015-03-01 11:51:55,817: WARNING/Worker-2] hello world
[2015-03-01 11:51:55,818: INFO/MainProcess] Task celery_client.
work[85eb0e88-cc6a-4c0a-a7e5-dcf8fabd810a] succeeded in
0.00152281699957s: None
```

We see our `hello world` message printed clearly! That's it, basically. You're left with a clean, purposeful code, which doesn't leak any kind of AMQP implementation detail on one side; and on the other side, you can trust that Celery packs a painfully major punch for the future and could tackle any kind of background processing scenario you might give it.

Let's sculpt our Pika-based scraper into Celery and hope to see how it cleans up our code. To do this, we only need to change the worker. The producer will be a simple call onto the worker, just as we did now.

Celery scraper

Let's start by rewriting our Pika scraper:

```
#!/usr/bin/env python
import requests
fromBeautifulSoup import BeautifulSoup
from celery import Celery

app = Celery('celery_pages', broker='amqp://guest@localhost//')

@app.task
def scrape(url):
print "-> Starting: [%s]" % (url)
    r = requests.get(url)
soup = BeautifulSoup(r.text)
print "-> Extracted: %s" % (soup.html.head.title)
print "-> Done: [%s]" % (url)
```

Wow. The code looks *a lot cleaner* and *a lot shorter*. Notice that we don't need to specify a lot of implementation detail in terms of RabbitMQ/AMQP—just a broker's location and a name of a queue. There's also no kind of job acknowledgement, because the Celery framework takes care of it all.

Celery scheduler

We still need to schedule jobs in. We start again by rewriting the previous Pika scraper:

```
#!/usr/bin/env python
import schedule
import time
from celery import Celery
```

```
fromcelery_scraper import scrape

app = Celery('celery_pages', broker='amqp://guest@localhost//')

urls = [ "http://ebay.to/1G163Lh" ]

def produce():
forurl in urls:
scrape.delay(url)
print("* Submitted: [%s]" % (url))

schedule.every(10).seconds.do(produce)

while True:
schedule.run_pending()
time.sleep(1)
```

Notice that we are reaching out to the `celery_scraper` module and pulling the `scrape` task to later use it in our `produce` callback.

We again initialize Celery by pointing it to a broker location (using the familiar AMQP URL) and queue.

Let's run and see how it all works together. Again, to highlight the process, we'll spawn a worker (or any number of these as needed); we'll also spawn the scheduler daemon in another terminal.

In production, you would automate such a thing using Foreman or an operating-system's `init` mechanisms such as `Upstart` or `Systemd` under Linux.

So, here we go; let's start the scraper via the Celery runner:

```
$ celery -A celery_scraper worker -loglevel=info
-------------- celery@jondot-mbp.local v3.1.17 (Cipater)
---- **** -----
--- * ***  * -- Darwin-13.4.0-x86_64-i386-64bit
-- * - **** ---
- ** ---------- [config]
- ** ---------- .> app:         celery_pages:0x102b6c9d0
- ** ---------- .> transport:   amqp://guest:**@localhost:5672//
- ** ---------- .> results:     disabled
- *** --- * --- .> concurrency: 4 (prefork)
-- ******* ----
--- ***** ----- [queues]
```

```
-------------- .> celery          exchange=celery(direct) key=celery


[tasks]
  . celery_scraper.scrape

[2015-03-01 12:02:22,015: INFO/MainProcess] Connected to amqp://
guest:**@127.0.0.1:5672//
[2015-03-01 12:02:22,030: INFO/MainProcess] mingle: searching for
neighbors
[2015-03-01 12:02:23,039: INFO/MainProcess] mingle: all alone
[2015-03-01 12:02:23,052: WARNING/MainProcess] celery@jondot-mbp.local
ready.
```

We see that it is up and running. And now, for the scheduler that will produce work, we have the following code:

```
$ python celery_scheduler.py
* Submitted: [http://ebay.to/1G163Lh]
* Submitted: [http://ebay.to/1G163Lh]
* Submitted: [http://ebay.to/1G163Lh]
```

Now, let's see what happened in the worker world:

```
[2015-03-01 12:02:25,051: INFO/MainProcess] Received task: celery_
scraper.scrape[1a5dc8c3-a6c8-461a-9431-1a7690652a77]
[2015-03-01 12:02:25,052: INFO/MainProcess] Received task: celery_
scraper.scrape[d28f137f-3811-4dcc-a51b-f5ece53dfb40]
[2015-03-01 12:02:25,053: WARNING/Worker-1] -> Starting: [http://ebay.
to/1G163Lh]
[2015-03-01 12:02:25,053: WARNING/Worker-2] -> Starting: [http://ebay.
to/1G163Lh]
[2015-03-01 12:02:25,065: INFO/Worker-1] Starting new HTTP connection
(1): ebay.to
[2015-03-01 12:02:25,065: INFO/Worker-2] Starting new HTTP connection
(1): ebay.to
[2015-03-01 12:02:25,517: INFO/Worker-2] Starting new HTTP connection
(1): www.ebay.com
[2015-03-01 12:02:25,519: INFO/Worker-1] Starting new HTTP connection
(1): www.ebay.com
[2015-03-01 12:02:26,952: WARNING/Worker-2] -> Extracted: <title>
Killer Rabbit of Death w Pointy Teeth Monty Python Blinking Red Eyes |
eBay </title>
[2015-03-01 12:02:26,953: WARNING/Worker-2] -> Done: [http://ebay.
to/1G163Lh]
```

```
[2015-03-01 12:02:26,953: INFO/MainProcess] Task celery_scraper.
scrape[1a5dc8c3-a6c8-461a-9431-1a7690652a77] succeeded in
1.901043879s: None
[2015-03-01 12:02:26,969: WARNING/Worker-1] -> Extracted: <title>
Killer Rabbit of Death w Pointy Teeth Monty Python Blinking Red Eyes |
eBay </title>
[2015-03-01 12:02:26,969: WARNING/Worker-1] -> Done: [http://ebay.
to/1G163Lh]
[2015-03-01 12:02:26,970: INFO/MainProcess] Task celery_scraper.
scrape[d28f137f-3811-4dcc-a51b-f5ece53dfb40] succeeded in
1.917738609s: None
[2015-03-01 12:02:31,226: INFO/MainProcess] Received task: celery_
scraper.scrape[721c2fbd-8625-45da-b690-822665961038]
[2015-03-01 12:02:31,227: WARNING/Worker-3] -> Starting: [http://ebay.
to/1G163Lh]
[2015-03-01 12:02:31,234: INFO/Worker-3] Starting new HTTP connection
(1): ebay.to
[2015-03-01 12:02:31,691: INFO/Worker-3] Starting new HTTP connection
(1): www.ebay.com
[2015-03-01 12:02:33,010: WARNING/Worker-3] -> Extracted: <title>
Killer Rabbit of Death w Pointy Teeth Monty Python Blinking Red Eyes |
eBay </title>
[2015-03-01 12:02:33,010: WARNING/Worker-3] -> Done: [http://ebay.
to/1G163Lh]
[2015-03-01 12:02:33,011: INFO/MainProcess] Task celery_scraper.
scrape[721c2fbd-8625-45da-b690-822665961038] succeeded in
1.784036434s: None
```

Our tasks are happily processed by our worker! To me this is awesome; Celery lets you write only the code that you need, resulting in a very good signal-to-noise ratio for you—the developer.

Exploring Celery

Just to give you an idea about the amount of work Celery does for you, let's breeze through a couple of other things that you can do with the help of Celery

Scheduling

Well, surprise! The scheduler that we have implemented is so common a pattern that the Celery framework already has a generic one in its ecosystem. Let's see how to use it:

```
fromcelery.schedules import crontab

CELERYBEAT_SCHEDULE = {
# Executes every Monday morning at 7:30 A.M
'add-every-monday-morning': {
'task': 'tasks.add',
'schedule': crontab(hour=7, minute=30, day_of_week=1),
'args': (16, 16),
},
}
```

You'll notice that we can specify a schedule here using a crontab-style schedule, and we can also mention the task and arguments that we want to schedule. This is quite simple and efficient.

We can then start the scheduling service called `beat`. We then start it with the following code:

```
$ celery -A proj beat
```

For more information regarding scheduling, see `http://docs.celeryproject.org/en/latest/userguide/periodic-tasks.html`.

HTTP hook tasks

Some tasks will look the same no matter what frameworks you implement them for. One of these is the task of integration between systems. A typical scenario would be your code against another system that you don't have access to its code, belongs to a different company, or is just written in another language.

The way to integrate this is with the help of a web hook. This means that the entity you want to integrate with will expose functionality via a simple HTTP call that you can make. A famous example by now would be Github's web hook, which allows you to integrate against events such, as commits, pull requests, and so on.

What you would do in the context of Celery is just use a boilerplate task. But before this, let's assume we have a simple route defined in Ruby on Rails (or any other platform, as long as it is radically different than ours for the sake of this example):

```
def hello(p)
res = {:status =>'success', :retval => "hello #{p}"}
render :json => res
end
```

This will simply return the `hello` string in a JSON data structure when called via HTTP (assuming we plug it into a controller and set up a route to it, but this is beyond the scope for now).

You can then try it out in the interactive Python shell:

```
>>>fromcelery.task.http import URL
>>>res = URL('http://example.com/hello').get_async("world")
```

This concludes our session into hooks.

Other Celery features

Celery sports additional powerful features, shown as follows:

- **Job Orchestration**: When you arrive at the point where you're looking at a complex mesh of jobs with interdependencies, Celery lets you describe this graph with simple functional building blocks, such as `group`, `map`, `mapstar`, and so on. It will then resolve these automatically for you, leaving you with a single result to wait for.
- **Task Routing**: This is the place where Celery lets you touch at core messaging semantics, such as fan-out, and apply them to your own tasks and workers. Most of the functionalities relate directly to the RabbitMQ back end.
- **Monitoring and Management**: Celery packs its own tooling for this. Although, RabbitMQ offers its own `rabbitmqctl` tool, Celery's tool offers insights into the Celery model, such as all states of tasks, queues, and more.
- **Advanced Worker Semantics**: Celery maintains a gossip network of all workers. This means that should you want, you could implement advanced mechanics, such as worker to worker introspection, job stealing logic, and shared worker state, among other things.

These just scratch the surface of what Celery packs. Since we cannot cover the entirety of the framework, we'll sadly stop here. But, feel free to continue exploring Celery. With this kind of framework, you can build very successful products and even a career on, so it is worth while to know it well.

Summary

To conclude this chapter, let's review what we have learned:

- **Python and the python ecosystem**: With regards to RabbitMQ, we have shown how easy it is to have the first step into the AMQP world.
- **Pika**: We made a deep dive into Pika, shown how simple it is to implement a common producer and consumer
- **Scraper project**: We have implemented a simple scraper backbone that you can use in your future projects, using just Pika. Simplicity is key here, because more often than not, people can over-complicate things.
- **Pika API**: We have covered the important surface of Pika's API and understood which scenarios they might come useful for.
- **Celery**: We have introduced the de-facto background processing framework for Python that relies on RabbitMQ primarily (as well as other backends)
- **Celery Scraper**: We have shown how we can easily, almost automatically migrate the "old" Scraper code into Celery, and seen how cleaner the code is.
- **Celery features**: We went over some other celery features, including scheduling and HTTP hooks. And saw that in some point, Celery even implements our own custom code off-the-shelf (scheduler) within the Scraper project.

After reading this chapter, I advise you to remember Celery and Pika and learn them well in this order.

In your day-to-day Python work, using Celery will feel as a second language and using such proper tooling for background jobs will give you the X-Factor against any other Python programmer.

Index

R

S

T

U

V

W

Y

Z

[PACKT] open source
PUBLISHING
community experience distilled

Thank you for buying
Mastering RabbitMQ

About Packt Publishing

Packt, pronounced 'packed', published its first book, *Mastering phpMyAdmin for Effective MySQL Management*, in April 2004, and subsequently continued to specialize in publishing highly focused books on specific technologies and solutions.

Our books and publications share the experiences of your fellow IT professionals in adapting and customizing today's systems, applications, and frameworks. Our solution-based books give you the knowledge and power to customize the software and technologies you're using to get the job done. Packt books are more specific and less general than the IT books you have seen in the past. Our unique business model allows us to bring you more focused information, giving you more of what you need to know, and less of what you don't.

Packt is a modern yet unique publishing company that focuses on producing quality, cutting-edge books for communities of developers, administrators, and newbies alike. For more information, please visit our website at `www.packtpub.com`.

About Packt Open Source

In 2010, Packt launched two new brands, Packt Open Source and Packt Enterprise, in order to continue its focus on specialization. This book is part of the Packt Open Source brand, home to books published on software built around open source licenses, and offering information to anybody from advanced developers to budding web designers. The Open Source brand also runs Packt's Open Source Royalty Scheme, by which Packt gives a royalty to each open source project about whose software a book is sold.

Writing for Packt

We welcome all inquiries from people who are interested in authoring. Book proposals should be sent to `author@packtpub.com`. If your book idea is still at an early stage and you would like to discuss it first before writing a formal book proposal, then please contact us; one of our commissioning editors will get in touch with you.

We're not just looking for published authors; if you have strong technical skills but no writing experience, our experienced editors can help you develop a writing career, or simply get some additional reward for your expertise.

[PACKT] open source
community experience distilled
PUBLISHING

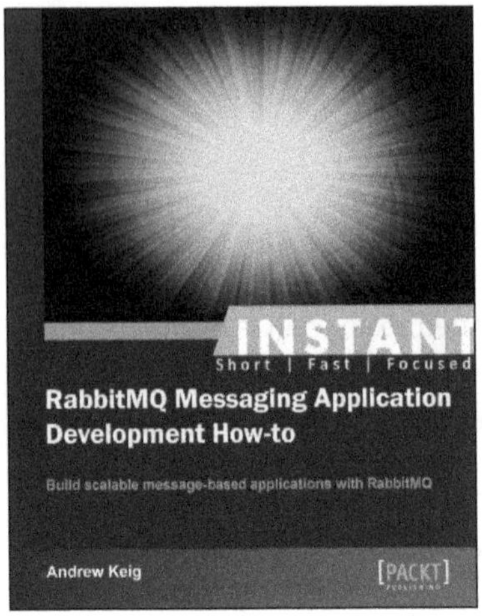

Instant RabbitMQ Messaging Application Development How-to

ISBN: 978-1-78216-574-3 Paperback: 54 pages

Build scalable message-based applications with RabbitMQ

1. Learn something new in an Instant! A short, fast, focused guide delivering immediate results.
2. Learn how to build message-based applications with RabbitMQ using a practical Node.js ecommerce example.
3. Implement various messaging patterns including asynchronous work queues, publish subscribe and topics.

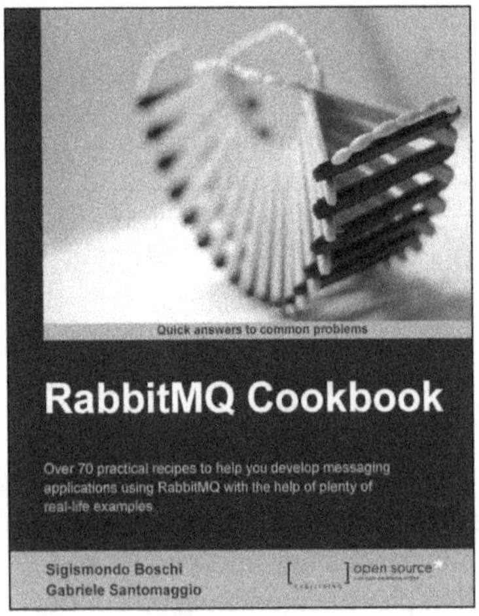

RabbitMQ Cookbook

ISBN: 978-1-84951-650-1 Paperback: 288 pages

Over 70 practical recipes to help you develop messaging applications using RabbitMQ with the help of plenty of real-life examples

1. Create scalable distributed applications with RabbitMQ.
2. Exploit RabbitMQ on both Web and mobile platforms.
3. Deploy message services on cloud computing platforms.

Please check **www.PacktPub.com** for information on our titles

[PACKT] open source
community experience distilled
PUBLISHING

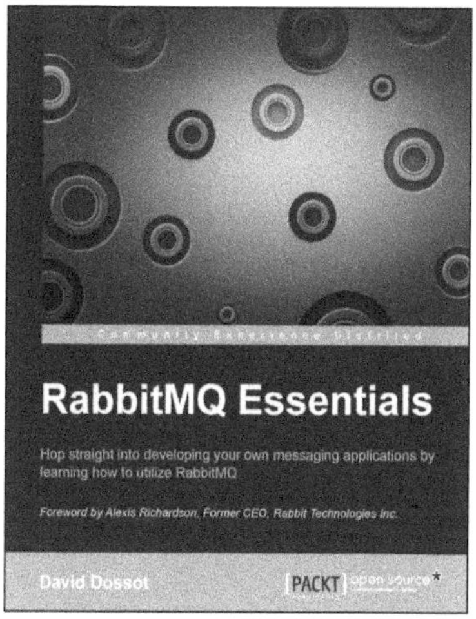

RabbitMQ Essentials

ISBN: 978-1-78398-320-9 Paperback: 182 pages

Hop straight into developing your own messaging applications by learning how to utilize RabbitMQ

1. Refresh your knowledge of the basics of message-orientated architecture and witness how powerful RabbitMQ can be when building your messaging applications.
2. Discover the strategies behind increasing the scalability and fault tolerance of your applications.
3. Gain a deep and practical understanding of RabbitMQ through the journey of Clever Coney Media, a fictitious company with real-world problems.

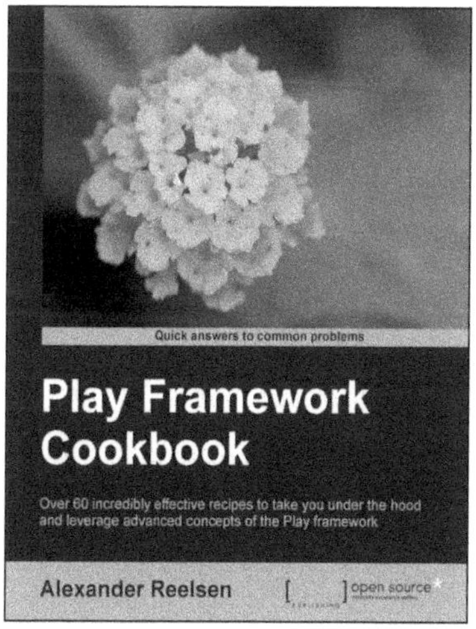

Play Framework Cookbook

ISBN: 978-1-84951-552-8 Paperback: 292 pages

Over 60 incredibly effective recipes to take you under the hood and leverage advanced concepts of the Play framework

1. Make your application more modular, by introducing you to the world of modules.
2. Keep your application up and running in production mode, from setup to monitoring it appropriately.
3. Integrate play applications into your CI environment.
4. Keep performance high by using caching.

Please check **www.PacktPub.com** for information on our titles

www.ingramcontent.com/pod-product-compliance
Lightning Source LLC
Chambersburg PA
CBHW061348210326
41598CB00035B/5922